'The witty and incisive spliced author, J. K. Gibson-Graham, has given us a superb tool for undoing the strangling grip of the ways we understand capitalism.'

Professor Donna Haraway, University of California, Santa Cruz

'After the fall of the Soviet Union, capitalism was said, from far left to far right, to be in the pink of health, never better, robust, seeping into every nook and cranny of global space, feeling strong, unified, and ready for action. Poor capitalism. J. K. Gibson-Graham's *The End of Capitalism (As We Knew It)* has sent capitalism back to its room. This courageous and brilliant book takes on all the current triumphalist discourses of capitalism's global spread. Borrowing from queer theory, feminism, and Marxism, Gibson-Graham rebuilds social theory and political economy around the notion of economic difference. She brings capitalism down a peg or two, claiming it to be one of several competitors for economic space, even if it has monopolized current narratives of political economy. Gibson-Graham has done us all an enormous service; she has shown, in graceful, witty, and pointed prose, that the battle to reclaim economic space in the name of equality, pleasure, and the end of exploitation starts with capitalism's representations.'

Jack Amariglio, Editor, Rethinking Marxism

'Ever wonder why "feminists have revolution now, while Marxists have to wait"? Feeling overwhelmed by "The Economy"? Tired of fabulous theories of Capitalism that make an anti-capitalist politics seem pointless, remote and grim? Paralyzing problems are banished by this dazzlingly lucid, creative and practical rethinking of class and economic transformation. I have wanted to read a strong, rigorous, feminist critique of political economy for a long time; I did not hope for one so generous in spirit, so dynamic in style, so inspiring in its implications. J. K. Gibson-Graham is a major theorist.'

Meaghan Morris, ARC Senior Fellow,
University of Technology, Sydney

'In this profoundly imaginative study, Gibson-Graham proposes an altogether new range of ways to break the paralysis that a reified concept of capitalism has cast over oppositional theory and politics. As fearless in conception as it is friendly in style, this is a book that evokes rare emotions: surprise and hope.'

Eve Kosofsky Sedgwick,
Newman Ivey White Professor of English, Duke University

for Steve and Rick

THE END OF CAPITALISM

(as we knew it)

A Feminist Critique
of Political Economy

J.K. Gibson-Graham

First published 1996
Reprinted 1997

Blackwell Publishers Inc
350 Main Street
Malden, Massachusetts 02148, USA

Blackwell Publishers Ltd
108 Cowley Road
Oxford OX4 1JF, UK

Library of Congress Cataloging in Publication Data
Gibson-Graham, J. K.
The end of capitalism (as we knew it): a feminist critique of political economy / J. K. Gibson-Graham.
p. cm.
Includes bibliographical references and index.
ISBN 1–55786–862–X — ISBN 1–55786–863–8 (pbk)
1. Capitalism. 2. Marxian economics. 3. Feminist economics.
I. Title.
HB501.G447 1996 95–52671
330.12'2 — dc20 CIP

British Library Cataloguing in Publication Data
A CIP catalogue record for this book is available from the British Library

Printed and bound in Great Britain
by Athenæum Press Ltd, Gateshead, Tyne & Wear

This book is printed on acid-free paper

Contents

Preface and Acknowledgments

Recently I was the commentator on a panel where I confronted the "problem" of this book, or a version of it, in the findings of a large collaborative research project. The panelists focused on the intersection of industrial restructuring and changing household work practices, and while their research was explicitly geared to exploring the reciprocal impacts and interactions of these two social domains, the principal relationship that emerged was the adaptation of households to changes in the industrial sector. "Adaptability" and "coping" were the general terms under which a remarkably diverse array of living situations and reactions to industrial change were subsumed.

The researchers hinted in passing at some points of tension and contradiction that stood out against the background of harmonization and adjustment: older workers who were having great difficulties adjusting to a 12-hour shift, destabilized gender identities stemming from a changing gender division of labor in two-earner households, increases in (noncapitalist and nonindustrial) economic activity among both shiftworkers and laid off workers. But while these problematic, contradictory, and complicating moments were acknowledged, their effects upon the industrial site were left unexplored. What was being produced was a narrative of local adaptability and accommodation, inadvertently establishing the dominance of global economic restructuring over local social and cultural life.

It was clear to me that the refusal to explore disharmony – the things that did not line up and fit in with industrial change – had led to an unwitting economism or productionism in the social representation that was being constructed. As the process to which everything else was adapting and adjusting, industrial restructuring was the central and determining dynamic in the local social setting, though this kind of deterministic analysis was the very thing the researchers had wanted to avoid. I tried carefully to suggest that in attempting to uncover "what was happening" in their local case study, their research project had become part of the process of restructuring itself; it had produced a language and an image of noncontradiction between capitalist workplace changes, changes in household practices and the constitution of gendered identities, and in this way it contributed to consolidating the affinities it represented.

In a very different discursive setting, I had recently encountered Eve Kosofsky Sedgwick's description of what she calls the "Christmas effect." To Sedgwick's mind what is so depressing about Christmas is the way all the institutions of society come together and speak "with one voice" (1993: 5): the Christian churches, of course, but also the state (which establishes school and national holidays), commerce, advertising, the media (revving up the Christmas frenzy and barking out the Christmas countdown), social events and domestic activities, "they all . . . line up with each other so neatly once a year, and the monolith so created is a thing one can come to view with unhappy eyes" (p. 6).

Sedgwick points to a similar monolithic formation in the realm of expectations about sexuality, where gender, object choice, sexual practices (including the privileging of certain organs and orifices), and "lifestyles" or life choices are expected to come together in predictable associations. This set of expectations, which counters and yet constrains the sexual experience of so many, is not just the occasion of seasonal distress. It is a source of lifelong oppression, a matter of survival, and a painful constrictor of sexual possibility, if not desire.

In my comments as a discussant I seemed to be chafing against a similarly constraining "Christmas effect" in the realm of social theory. The researchers had set out to produce a rich and differentiated set of stories about industrial and community change, but they ended up showing how households and communities accommodated to changes in the industrial sector. In their papers things not only *lined up with* but *revolved around* industry, producing a unified social representation centered on a capitalist economy (the sort of thing that's called a "capitalist society" in both everyday and academic discussion).

But Sedgwick's questions about Christmas, the family, and sexuality suggested the possibility of other kinds of social representations: "What if . . . there were a practice of valuing the ways in which meanings and institutions can be at loose ends with each other? What if the richest junctures weren't the ones where *everything means the same thing?*" (1993: 6). For this research project following Sedgwick's suggestions might mean that unstable gender identities, inabilities to adapt to the new shiftwork schedule, and noncapitalist economic activities should be emphasized rather than swept under the rug. The vision of households, subjects, and capitalist industry operating in harmony (and in fact coming together in a new phase of capitalist hegemony) might be replaced by alternative social representations in which noncapitalist economic practices proliferated, gender identities were renegotiated, and political subjects actively resisted industrial restructuring, thereby influencing its course.

More generally, Sedgwick's vision suggests the possibility of representing societies and economies as nonhegemonic formations. What if we were to depict social existence at loose ends with itself, in Sedgwick's terms, rather than producing social representations in which everything is part of the same complex and therefore ultimately "means the same thing" (e.g., capitalist hegemony)? What might be the advantages of representing a rich and prolific disarray?

I was particularly attuned to these problems and possibilities because I had myself been a producer, in my earlier work as a political economist, of representations of capitalist hegemony. As a member of a large and loosely connected group of political economic theorists who were interested in what had happened to capitalist economies following on the economic crisis of the 1970s, I had engaged in theorizing the ways in which industrial production, enterprises, forms of consumption, state regulation, business culture, and the realm of ideas and politics all seemed to undergo a change in the 1970s and 80s from one hegemonic configuration to another. It didn't matter that I was very interested in the differences between industries or that I did not see industrial change – even widespread change – as emanating from or reflecting a macrologic of "the economy." I was still representing a world in which economy, polity, culture, and subjectivity reinforced each other and wore a capitalist face. Chasing the illusion that I was understanding the world in order to change it, I was running in a well-worn track, and had only to cast a glance over my shoulder to see, as the product of my analysis, "capitalist society" even more substantial and definitive than when I began.

In those exciting early days I had yet to take seriously the "performativity" of social representations – in other words, the ways in which they are implicated in the worlds they ostensibly represent. I was still trying to capture "what was happening out there," like the researchers on the panel. I wasn't thinking about the social representation I was creating as *constitutive* of the world in which I would have to live. Yet the image of global capitalism that I was producing was actively participating in consolidating a new phase of capitalist hegemony.[1] Over a period of years this became increasingly clear to me and increasingly distressing.

My situation resembled that of the many other social theorists for whom the "object of critique" has become a perennial and consequential theoretical issue. When theorists depict patriarchy, or racism, or compulsory heterosexuality, or capitalist hegemony they are not only delineating a formation they hope to see destabilized or replaced. They are also generating a representation of the social world and endowing it with performative force. To the extent that this representation becomes influential it may contribute to the hegemony of a "hegemonic formation"; and it will undoubtedly influence people's ideas about the possibilities of difference and change, including the potential for successful political interventions.[2]

Perhaps it is partly for this reason that many social theorists have taken to theorizing a hegemonic formation in the field of discourse (heteronormativity, for instance, or a binary gender hierarchy) while representing the social field as unruly and diverse.[3] A good example can be found in Eve Sedgwick's opening chapter to *Epistemology of the Closet* where she counterposes to a heteronormative discourse of sexuality the "obviousness"[4] of the great and existing diversity of people's relations to sex. In a similar fashion, bell hooks sets a dominant phallocentric discourse of black masculinity (and black racial identity) against the diverse social field of black masculinities and gender relations.[5]

Like many political economists I had heretofore theorized the US social formation and "the global economy" as sites of capitalist dominance, a dominance located squarely in the social (or economic) field. But a theoretical option now presented itself, one that could make a (revolutionary)

1 When I heard union leaders exhorting their memberships to accept the realities of the new global economy and act accordingly to maintain their share of the pie, I felt in part responsible for the note of inevitability in their voices. As Fred Block has pointed out, "social theory plays an indispensable role in providing us with a roadmap to our social environment" (1990: 2). The kind of social theory I was producing mapped a terrain that was structured and governed by global capitalism, and that offered only a few highly constrained political options.

2 A feeling of hopelessness is perhaps the most extreme and at the same time most familiar political sentiment in the face of a massive or monolithic patriarchy, racism, or capitalism.

3 Of course, this is a strategy that has its failures and problems as well as its strengths and successes. Just because something is discursive doesn't mean it's not monolithic or intractable, as Butler points out in her discussion of Irigaray's universalist and crosscultural construction of phallogocentrism (1990: 13) and her criticism of the way in which female abjection is sometimes treated as a founding structure in the "symbolic" domain: "Is this structure of feminine repudiation not *reenforced* by the very theory which claims that the structure is somehow prior to any given social organization, and as such resists social transformation?" (1995: 19).

4 An obviousness itself presumably constituted by nonhegemonic or marginal discourses.

5 See the chapter in *Black Looks* on black masculinity.

difference: to depict economic discourse as hegemonized while rendering the social world as economically differentiated and complex. It was possible, I realized, and potentially productive to understand capitalist hegemony as a (dominant) discourse rather than as a social articulation or structure. Thus one might represent economic practice as comprising a rich diversity of capitalist and noncapitalist activities and argue that the noncapitalist ones had until now been relatively "invisible" because the concepts and discourses that could make them "visible" have themselves been marginalized and suppressed.

Now we have arrived at the present, the moment of the writing (or actually the completion) of this book. The book has been written about, and against, discourses of capitalist hegemony. It attempts to clear a discursive space for the emergence and development of hitherto suppressed discourses of economic diversity, in the hope of contributing to an anticapitalist politics of economic invention.

Becoming able to envision and ultimately to write this book has involved for me the most profound transformations both in my intellectual work and in my relation to that work. These transformations extend to, or perhaps begin with, my personal identity. For it was only in the summer of 1992 that J. K. Gibson-Graham was born (in a dormitory room at Rutgers University where a feminist conference was taking place): Following in the steps of many women writers who have played with authorship and naming, we (Katherine Gibson and Julie Graham) became in that moment a single writing persona.[6] We had been working, thinking, and writing together for over fifteen years since undertaking a joint project on New England plant closings during our first year in graduate school. And it had become important to subvert in a practical fashion the myriad hierarchies of value and power that (in shifting and complex ways) structured our relationship, negotiated as it was across differences of nationality, age, appearance, academic training, family status, personality and experience, to name just a few.

[6] The example that seems closest to our own situation is that of the early twentieth-century Australian writers Marjorie Barnard and Flora Eldershaw whose writing persona, M. Barnard Eldershaw, was the author of a number of historical and contemporary novels. As a nationalist and feminist critic of the English literary canon and the importance of the Great Author, M. Barnard Eldershaw established a practice of, and a role model for, collective writing. One of the characters in her novel *Plaque with Laurel* argues for a literary community of complementary strengths and weaknesses: "Don't you think that perhaps we do make a whole between us, even if each one only contributes a little? It doesn't mean that there is no pattern because we can't see it. We might make up for one another. In the end somebody's plus fits into somebody's minus. Like a jigsaw, you know. There's a whole, but it doesn't belong to anyone. We share it" (Eldershaw 1937: 295).

Becoming JKGG liberated us from established notions about which of us was the "writer," the "researcher," the "theorist," the "creative thinker." It allowed us to celebrate and build upon our differences in ways that we had not previously done.[7] (That this combined persona has its own instabilities can be seen in our use of the first person, which sometimes appears as "I" and sometimes as "we".)

Our transformed relation to ourselves and our work has manifested itself in a much more adventurous approach to reading, writing, and the practice of research.[8] In particular we have increasingly ventured outside our disciplinary boundaries and into fields other than political economy and geography (where we received our training under the lavish mentoring of Don Shakow, Ron Horvath, Bennett Harrison, and Bob Ross.) Our tendency to stray and migrate has been fostered by our encounter with anti-essentialist Marxism as it has developed in the journal *Rethinking Marxism* and in the work of Steve Resnick and Rick Wolff, Jack Amariglio and David Ruccio, and other members of the Association for Economic and Social Analysis (AESA) – Enid Arvidson, Usha Rao Banerjee, George DeMartino, Jonathan Diskin, Becky Forest, Harriet Fraad, Rob Garnett, Janet Hotch, Susan Jahoda, Ric McIntyre, Bruce Norton, Luis Saez, Blair Sandler, Amy Silverstein, Jackie Southern, Kevin St. Martin, Marjolein van der Veen, Peter Wissoker, and others too numerous to mention.

The heady adventure of "rethinking Marxism" has encouraged us to draw upon the other forms of social theory that are currently experiencing an explosion of creativity and growth, in particular poststructuralist feminist and queer theory, to facilitate our specific projects of rethinking. Here we have been particularly influenced and enabled by the work of Michèle Barrett, Elizabeth Grosz, Jonathan Goldberg, Sharon Marcus, Meaghan Morris, Chantal Mouffe, and Eve Kosofsky Sedgwick, and by our intense and enriching friendships with Jack Amariglio and Michael Moon.

If intellectual pleasure-seeking has characterized the period in our lives during which this book has been written, we should not fail to mention

[7] It also had some unpredictable effects on our ways of working together. Previously our joint papers had been laboriously produced as joint products, with each of us doing a nearly equivalent amount and type of work on each paper. As JKGG we found ourselves still collaborating intellectually but adopting a more individualized writing process, with one person operating as the reader/consultant and the other as the writer of each chapter or paper.

[8] In the eyes of some of our social science and Marxist associates we have in fact undergone a transmutation from theory saints (faithful and devoted to the truth of the world) to theory sluts (good time girls who think around): if you're not with the theory you love, love the theory you're with.

some of the non-intellectual pleasures that have attended its emergence. Over the past six years we have frequently overcome the tyranny of distance that separates us by meeting for work retreats (somewhat misleadingly called "nunneries"). In these visits, free from the claims of jobs and family, we have luxuriated in the pleasures of single-minded concentration on thinking, talking and writing, interspersed with sumptuous meals, chocolate-covered Macadamia nuts, and outdoor activities in beautiful settings – Venus Bay near Melbourne, the islands of Oahu and Hawaii, the Black Hills of South Dakota. In many of these places dear friends and family members have offered accommodation and undemanding service: Beverly and George in Jackson Hole, Nat and Bill in Portola Valley, Judy and Jack in Cape Cod, Lillian and Peter in Balgowlah, Andrea and Peter in the Adirondacks.

Some of the most rewarding and demanding experiences that have pushed this project along have been the opportunities to interact on an extended basis with faculty and graduate students at a range of institutions. We would like to acknowledge the financial assistance, professional support, and intellectual challenges offered by Gordon Clark and a Faculty of Arts Visiting Fellowship at Monash University; the Center for Critical Analysis of Contemporary Culture at Rutgers University; Allen Hunter, Erik Olin Wright and the Havens Center at the University of Wisconsin–Madison; the Department of Geography at Penn State; Dick Bedford and the Department of Geography at the University of Waikato; Wallace Clement and Rianne Mahon and the Institute of Political Economy at Carleton University; and the Centre for Women's Studies at Monash University.

A number of people read the entire book in manuscript and made invaluable comments contributing to its final revision. These stalwart souls (and close friends) cannot escape without at least a little of the blame for the finished product: Jillana Enteen, Neil Talbot, David Tait, Marjolein van der Veen, we appreciate your perseverance and gentleness. David Ruccio spent an entire vacation generating helpful suggestions without once making us feel that we should have written a different book.[9]

Others who have provided intellectual nourishment and personal support, whether or not they were involved in or sympathetic to this particular project, include our friends and colleagues Caroline Alcorso, Laurie Brown, Dick Bryan, Marta Calas, Jenny Cameron, Lauren Costello, Robyn Cross, Nancy Duncan, Glen Elder, Bob Fagan, Ruth Fincher,

[9] Many of our own conceptions of the book originated with these readers, who variously characterized it as "wishful rethinking" (Talbot), "provocative and optimistic" (Ruccio), a "space of seduction" (Tait), and "a force to contend with" (Enteen).

Nancy Folbre, Pat Greenfield, Jim Hafner, Ginny Haller, Gay Hawkins, Carol Heim, Phil Hirsch, Ruth and Richard Hooke, Richie Howitt, Sophie Inwald, Jane Jacobs, John Paul Jones, Cindi Katz, Graham Larcombe, Wendy Larner, Margaret Lee, Eric Levitson, Sharon Livesey, Beth Loffreda, Robyn Longhurst, Rose Lucas, Suzanne MacKenzie, Ann Markusen, Peter McEvoy, Gabrielle Meagher, Andrew Metcalfe, Peter Murphy, Heidi Nast, Eve Oishi, Phil and Libby O'Neill, Richard Peet, Rosemary Pringle, Bruce Robbins, Romaine Rutnam, Lisa Saunders, Louisa Schein, Maaria Seppänen, Linda Smircich, Neil Smith, Paul Smith, Ed Soja, Ulf Strohmayer, Noël Sturgeon, Andrew Vickers, and Sophie Watson.

John Davey of Blackwell was a wonderfully encouraging and helpful editor throughout the process of getting the book written and produced. Luis Saez helped us with the extensive library work, indefatigably finding and hauling books. Thanks are due to Tony Grahame for his editorial amendments and the courteous spirit in which they were made.

Lastly, we have been able to sustain our friendship and the long-term working relationship that produced this book in part because of the unstinting love and support of several extended and domestic communities: the members of AESA, the residents of Cooleyville, Helen, Alfie, Josh, and David, Daniel and Lillian Tait.

The author and publishers wish to thank the following for permission to use portions of previously published copyright material:

"Rethinking Class in Industrial Geography: Creating a Space for an Alternative Politics of Class," *Economic Geography*, vol. 68, no. 2, (1992), pp. 109–27.

"Post-Fordism as Politics," first published in *Environment and Planning D: Society and Space*, vol. 10 (1992), pp. 393–410.

"'Hewers of Cake and Drawers of Tea': Women, Industrial Restructuring, and Class Processes on the Coalfields of Central Queensland," *Rethinking Marxism*, vol. 5, no. 4 (Winter 1992), pp. 29–56.

"Waiting for the Revolution, or How to Smash Capitalism While Working at Home in Your Spare Time, *Rethinking Marxism*," vol. 6, no. 2 (Summer 1993), pp. 10–24.

"Haunting Capitalism: Ghosts on a Blackboard," *Rethinking Marxism* (forthcoming).

"Querying Globalization," *Rethinking Marxism* (forthcoming).

1

Strategies

Understanding capitalism has always been a project of the left, especially within the Marxian tradition. There, where knowledges of "capitalism" arguably originated, theory is accorded an explicit social role. From Marx to Lenin to the neo-Marxists of the post-World War II period, theorists have understood their work as contributing – whether proximately or distantly – to anticapitalist projects of political action. In this sense economic theory has related to politics as a subordinate and a servant: we understand the world in order to change it.

Given the avowed servitude of left theory to left political action it is ironic (though not surprising) that understandings and images of capitalism can quite readily be viewed as contributing to a *crisis* in left politics. Indeed, and this is the argument we wish to make in this book, the project of understanding the beast has itself produced a beast, or even a bestiary; and the process of producing knowledge in service to politics has estranged rather than united understanding and action. Bringing these together again, or allowing them to touch in different ways, is one of our motivating aspirations.

"Capitalism" occupies a special and privileged place in the language of social representation. References to "capitalist society" are a commonplace of left and even mainstream social description, as are references – to the market, to the global economy, to postindustrial society – in which an unnamed capitalism is implicitly invoked as the defining and unifying moment of a complex economic and social formation. Just as the economic system in eastern Europe used confidently to be described as communist or socialist, so a general confidence in economic classification characterizes representations of an increasingly capitalist world system. But what might be seen as the grounds of this confidence, if we put aside notions of "reality" as the authentic origin of its representations?

Why might it seem problematic to say that the United States is a Christian nation, or a heterosexual one, despite the widespread belief that Christianity and heterosexuality are dominant or majority practices in their respective domains, while at the same time it seems legitimate and indeed "accurate" to say that the US is a capitalist country?[1] What is it about the former expressions, and their critical history, that makes them visible as "regulatory fictions,"[2] ways of erasing or obscuring difference, while the latter is seen as accurate representation? Why, moreover, have embracing and holistic expressions for social structure like patriarchy fallen into relative disuse among feminist theorists (see Pringle 1995; Barrett and Phillips 1992) while similar conceptions of capitalism as a system or "structure of power" are still prevalent and resilient? These sorts of questions, by virtue of their scarcity and scant claims to legitimacy, have provided us a motive for this book.[3]

The End of Capitalism (As We Knew It) problematizes "capitalism" as an economic and social descriptor.[4] Scrutinizing what might be seen as throwaway uses of the term – passing references, for example, to the capitalist system or to global capitalism – as well as systematic and deliberate attempts to represent capitalism as a central and organizing feature of modern social experience, the book selectively traces the discursive origins of a widespread understanding: that capitalism is the hegemonic, or even the only, present form of economy and that it will continue to be so in the proximate future. It follows from

1 For one thing, an ambiguity exists in the former instances (between, for example, the reference to a population and its heterosexual practices, and the reference to a regime of compulsory heterosexuality) that does not exist in the latter. This suggests that the "dominance" of capitalism might itself be undermined by representing capitalism as a particular set of activities practiced by individuals.

2 Butler (1990) uses this term with respect to the "fiction" of binary gender and its regulatory function as a support for compulsory heterosexuality. No matter how much the New(t) Right in the US wants to impose the "truth" of a Christian heterosexual nation, this fiction is actually the focus of considerable contention.

3 The list of questions could be extended. How is it, for example, that "woman" as a natural or extradiscursive category has increasingly receded from view, yet "capitalism" retains its status as a given of social description? The answer that presents itself to us has to do with the feminist politics of representation and the vexed problem of gender (and other forms of personal) identity. The question of social identity has not been so extensively vexed (despite the efforts of Laclau and Mouffe, among others) but is perhaps ripe for the vexing.

Many people have observed that the economic and social realms are sometimes accorded the status of an extratextual reality. Butler notes, for example, that the domain of the social is often seen as "given or already constituted." She suggests a reinfusion of what she calls "ideality," with its implications of "possibility" and "transformability," into feminist representations of the social (1995: 19–20).

this prevalent though not ubiquitous view that noncapitalist economic sites, if they exist at all, must inhabit the social margins; and, as a corollary, that deliberate attempts to develop noncapitalist economic practices and institutions must take place in the social interstices, in the realm of experiment, or in a visionary space of revolutionary social replacement.

Representations of capitalism are a potent constituent of the anticapitalist imagination, providing images of what is to be resisted and changed as well as intimations of the strategies, techniques, and possibilities of changing it. For this reason, depictions of "capitalist hegemony" deserve a particularly skeptical reading. For in the vicinity of these representations, the very idea of a noncapitalist economy takes the shape of an unlikelihood or even an impossibility. It becomes difficult to entertain a vision of the prevalence and vitality of noncapitalist economic forms, or of daily or partial replacements of capitalism by noncapitalist economic practices, or of capitalist retreats and reversals. In this sense, "capitalist hegemony" operates not only as a constituent of, but also as a brake upon, the anticapitalist imagination.[5] What difference might it make to release that brake and allow an anticapitalist economic imaginary to develop unrestricted?[6] If we were to dissolve the image that looms in the economic foreground, what shadowy economic forms might come forward? In these questions we can identify the broad outlines of our project: to discover or create a world of economic difference, and to populate that world with exotic creatures that become, upon inspection, quite local and familiar (not to mention familiar beings that are not what they seem).

The discursive artifact we call "capitalist hegemony" is a complex

[4] Though we refer on almost every page of this book to capitalism, we find ourselves loath to define it, since this would involve choosing among a wide variety of existing definitions (any one of which could be seen as our "target") or specifying out of context a formation that we wish to understand as contextually defined. One familiar Marxist definition, however, involves a vision of capitalism as a system of generalized commodity production structured by (industrial) forces of production and exploitative production relations between capital and labor. Workers, bereft of means of production, sell their labor power for wages and participate in the labor process under capitalist control. Their surplus labor is appropriated by capitalists as surplus value. The capitalist mode of production is animated by the twin imperatives of enterprise competition and capital accumulation which together account for the dynamic tendencies of capitalism to expand and to undergo recurring episodes of crisis.

[5] Which we hesitate to call "socialist" because of the emptiness of the term in a context where the meaning of capitalism is called into question. Conversely, of course, the "death" of socialism is one of the things that has made it possible to question and rethink capitalism (since each has largely been defined in opposition to the other).

[6] The metaphor of the brake is drawn from Haraway (1991: 41–2).

effect of a wide variety of discursive and nondiscursive conditions.[7] In this book we focus on the practices and preoccupations of discourse, tracing some of the different, even incompatible, representations of capitalism that can be collated within this fictive summary representation. These depictions have their origins in the diverse traditions of Marxism, classical and contemporary political economy, academic social science, modern historiography, popular economic and social thought, western philosophy and metaphysics, indeed, in an endless array of texts, traditions and infrastructures of meaning. In the chapters that follow, only a few of these are examined for the ways in which they have sustained a vision of capitalism as the dominant form of economy, or have contributed to the possibility or durability of such a vision. But the point should emerge none the less clearly: the virtually unquestioned dominance of capitalism can be seen as a complex product of a variety of discursive commitments, including but not limited to organicist social conceptions, heroic historical narratives, evolutionary scenarios of social development, and essentialist, phallocentric, or binary patterns of thinking. It is through these discursive figurings and alignments that capitalism is constituted as large, powerful, persistent, active, expansive, progressive, dynamic, transformative; embracing, penetrating, disciplining, colonizing, constraining; systemic, self-reproducing, rational, lawful, self-rectifying; organized and organizing, centered and centering; originating, creative, protean; victorious and ascendant; self-identical, self-expressive, full, definite, real, positive, and capable of conferring identity and meaning.[8]

The argument revisited: it is the way capitalism has been "thought" that has made it so difficult for people to imagine its supersession.[9] It

7 The latter including, among other things, working-class struggles and the forms of their successes and defeats. To take another example, the technologies of communication and replication that are used to trumpet the triumph of global capitalism are themselves nondiscursive conditions of "capitalist hegemony."

8 This list of qualities should not be seen as exhaustive. Indeed one could certainly construct a list of equal length that enumerated capitalism's weaknesses and "negative" characteristics: for example, images of capitalism as crisis-ridden, self-destructive, anarchic, requiring regulation, fatally compromised by internal contradictions, unsustainable, tending to undermine its own conditions of existence. That these opposing lists do not negate (or even substantially compromise) each other is one of the premises of this discussion. (In fact, "weaknesses" or problems of capitalism are often consonant with, and constitutive of, its perceived hegemony and autonomy as an economic system.)

9 Except, of course, as the product of evolutionary necessity or the millennial project of a revolutionary collective subject. At this moment on the left, when these two familiar ways of thinking capitalist supersession are in disrepair and disrepute, there are few ways of conceptualizing the replacement of capitalism by noncapitalism that we find persuasive.

is therefore the ways in which capitalism is known that we wish to delegitimize and displace. The process is one of unearthing, of bringing to light images and habits of understanding that constitute "hegemonic capitalism" at the intersection of a set of representations. This we see as a first step toward theorizing capitalism without representing dominance as a natural and inevitable feature of its being. At the same time, we hope to foster conditions under which the economy might become less subject to definitional closure. If it were possible to inhabit a heterogeneous and open-ended economic space whose identity was not fixed or singular (the space potentially to be vacated by a capitalism that is necessarily and naturally hegemonic) then a vision of noncapitalist economic practices as existing and widespread might be able to be born; and in the context of such a vision, a new anticapitalist politics might emerge, a noncapitalist politics of class (whatever that may mean) might take root and flourish. A long shot perhaps but one worth pursuing.

In this introduction we touch upon the various discursive appearances of capitalism that are given different or more detailed treatment later in the book. The introduction serves to convene them, and in bringing them together to make them susceptible to a single critique. As the prelude to and precondition of a theory of "economic difference," the critique of economic sameness (or of essentialism, to invoke a freighted synonymy) attempts to liberate a heterospace of both capitalist and noncapitalist economic existence. Here, as throughout the book, we draw upon the strategies of postmodern Marxism and poststructuralist feminism to enable both criticism and re-imagination. Somewhat diffidently and rudimentarily, we also take up the challenge of concretely specifying different economic practices that can be seen to inhabit a space of economic diversity, or that might be called into being to fulfill its promises of plenitude and potentiation. Together, the critical project of undermining prevalent practices of capitalist representation, and the more arduous project of generating a discourse of economic difference, constitute the unevenly distributed burden of this book.[10]

Strategy 1: Constructing the straw man

Capitalism's hegemony emerges and is naturalized in the space of its overlapping and intersecting appearances – as the earthly kingdom of modern industrial society; the heroic transformative agent of development/mod-

[10] In this book we give some glimpses of the noncapitalist class relations that inform our anticapitalist imaginary. Extended explorations of these class processes and positions are provided in our co-edited collection which is tentatively entitled *Class: The Next Postmodern Frontier* (in progress).

ernization; a unitary, structured and self-reproducing economic system; a protean body with an (infinite?) repertory of viable states; a matrix of flows that integrates the world of objects and signs; the phallus that structures social space and confers meaning upon social practices and positions (these as well as other representations are explored in later chapters.) Each of these figurings tends to position capitalism – with respect both to other specific types of economy and to the general social space of economic difference – as the dominant economic form. In other words not only is capitalism in itself triumphant, encompassing, penetrating, expansive (and so on), but by virtue of these "internal" capitalist qualities, other forms of economy are vanquished, marginalized, violated, restricted. Different as they may be from one another, they are united by their common existence as subordinated and inferior states of economic being. In this sense, we may speak of the relation of capitalism to noncapitalism in the terms of the familiar binary structure in which the first term is constituted as positivity and fullness and the second term as negativity or lack.

When we say that most economic discourse is "capitalocentric," we mean that other forms of economy (not to mention noneconomic aspects of social life) are often understood primarily with reference to capitalism: as being fundamentally the same as (or modeled upon) capitalism, or as being deficient or substandard imitations; as being opposite to capitalism; as being the complement of capitalism;[11] as existing in capitalism's space or orbit. Thus noncapitalist practices like self-employment may be seen as taking place *within* capitalism, which is understood as an embracing structure or system. Or noncapitalist activity may be elided, as when "commodification" is invoked as a metonym for capitalist expansion.[12] Noncapitalist economic forms may

[11] We are indebted for this definition to the conceptions of phallocentrism of Grosz (1990) and Irigaray.

[12] Despite the general recognition that slave, communal, family, independent and other production relations are all compatible with commodity production, that is, production of goods and services for a market, the commodity is often uniquely associated with capitalism (perhaps because of the prevalent definition of capitalism as involving "generalized" commodity production, referring to the existence of labor power as a commodity). Laclau and Mouffe depict the process of capitalist expansion over the post-World War II period in terms of commodity relations: "this 'commodification' of social life destroyed previous social relations, replacing them with commodity relations through which the logic of capitalist accumulation penetrated into increasingly numerous spheres . . . There is practically no domain of individual or collective life which escapes capitalist relations" (1985: 161). Note here the language of destruction, penetration, capture, replacement, invasion, and the sense that these processes are driven by a logic (in other words they are the phenomenal expressions of an underlying essence). See also chapter 6 on globalization.

be located in "peripheral" countries that lack the fullness and completeness of capitalist "development."[13] Noncapitalism is found in the household, the place of woman, related to capitalism through service and complementarity. Noncapitalism is the before or the after of capitalism: it appears as a precapitalist mode of production (identified by its fate of inevitable supersession); it appears as socialism, for which capitalism is both the negative and the positive precondition.

Capitalism's others fail to measure up to it as the true form of economy: its feminized other, the household economy, may be seen to lack its efficiency and rationality; its humane other, socialism, may be seen to lack its productivity; other forms of economy lack its global extensiveness, or its inherent tendency to dominance and expansion. No other form displays its systemic qualities or its capacity for self-reproduction (indeed projects of theorizing noncapitalism frequently founder upon the analogical imperative of representing an economic totality, complete with crisis dynamics, logics and "laws of motion"). Thus despite their ostensible variety, noncapitalist forms of economy often present themselves as a homogeneous insufficiency rather than as positive and differentiated others.

To account for the demotion and devaluation of noncapitalism[14] we must invoke the constitutive or performative force of economic representation. For depictions of capitalism – whether prevalent and persistent or rare and deliquescent – position noncapitalism in relations of subsumption, containment, supersession, replication, opposition and complementarity to capitalism as the quintessential economic form.[15] To take a few examples from a list that is potentially infinite:

(1) Capitalism appears as the "hero" of the industrial development narrative, the inaugural subject of "history," the bearer of the future, of modernity, of universality. Powerful, generative, uniquely sufficient to

[13] "Development" is not understood here as a process but in another of its meanings as the quintessential form of western society.

[14] Here and throughout, when we refer to noncapitalism, we mean noncapitalist forms of economy, unless otherwise noted.

[15] Of course some of the most famous and seminal representations of capitalism can be found in the *Communist Manifesto*, which came to life as one of the founding documents of a revolutionary political tradition. That the *Manifesto* – and the vision that animated it – functioned powerfully to motivate successful workers' movements is something we do not wish to deny; but the image of two classes locked in struggle has in our view now become an obstacle to, rather than a positive force for, anticapitalist political endeavors. It is difficult for us – and we believe for others – to identify with this image today, though it may still resonate with many.

the task of social transformation,[16] capitalism liberates humanity from the struggle with nature. (In its corresponding role as antihero, capitalist development bears the primary responsibility for underdevelopment and environmental degradation.)

(2) Capitalism is enshrined at the pinnacle of social evolution. There it brings – or comes together with – the end of scarcity, of traditional social distinctions, of ignorance and superstition, of antidemocratic or primitive political forms (this is the famous social countenance of modernization).[17] The earthly kingdom of modernism is built upon a capitalist economic foundation.

(3) Capitalism exists as a unified system or body, bounded, hierarchically ordered, vitalized by a growth imperative, and governed by a telos of reproduction. Integrated, homogeneous, coextensive with the space of the social, capitalism is the unitary "economy" addressed by macroeconomic policy and regulation. Though it is prone to crises (diseases), it is also capable of recovery or restoration.

(4) Capitalism is an architecture or structure of power, which is conferred by ownership and by managerial or financial control. Capitalist exploitation is thus an aspect or effect of domination, and firm size and spatial scope an index of power (quintessentially embodied in the multinational corporation).

(5) Capitalism is the phallus or "master term" within a system of social differentiation. Capitalist industrialization grounds the distinction between core (the developed world) and periphery (the so-called Third World). It defines the household as the space of "consumption" (of capitalist commodities) and of "reproduction" (of the capitalist workforce) rather than as a space of noncapitalist production and consumption.

Capitalism confers meaning upon subjects and other social sites in relation to itself, as the contents of its container, laid out upon its grid, identified and valued with respect to its definitive being. Complexly generated social processes of commodification, urbanization, internationalization,

[16] Anderson depicts capitalism in familiar terms as a relentless transformative force, one that "tears down every ancestral confinement and claustral tradition in an immense clearing operation of cultural and customary debris across the globe" (1988: 318). In a similar vein Spivak evokes capitalism's agency in service of its own imperatives: "To minimize circulation time, industrial capitalism needed to establish due process, and such civilizing instruments as railways, postal services, and a uniformly graded system of education" (1988b: 90).

[17] Acknowledging not only capitalism's agency but its extraordinary creativity and universalizing reach, Haraway invokes a feminist political imaginary by calling for "an emerging system of world order analogous in its novelty and scope to that created by industrial capitalism" (1991: 203). The earthly kingdom of capitalism can only be replaced by its likeness.

proletarianization are viewed as aspects of capitalism's self-realization.

(6) Capitalism's visage is plastic and malleable, its trajectory protean and inventive.[18] It undergoes periodic crises and emerges regenerated in novel manifestations (thus Fordism is succeeded by post-Fordism, organized by disorganized capitalism, competitive by monopoly or global capitalism).

(7) Ultimately capitalism is unfettered by local attachments, labor unions, or national-level regulation. The global (capitalist) economy is the new realm of the absolute, the not contingent, from which social possibility is dictated or by which it is constrained. In this formulation economic determinism is reborn and relocated, transferred from its traditional home in the "economic base" to the international space of the pure economy (the domain of the global finance sector and of the all powerful multinational corporation).

(8) It is but one step from global hegemony to capital as absolute presence: "a fractal attractor whose operational arena is immediately coextensive with the social field" (Massumi 1993: 132), "an enormous . . . monetary mass that circulates through foreign exchange and across borders," "a worldwide axiomatic" (Deleuze and Guattari 1987: 453) engaged in "the relentless saturation of any remaining voids and empty places" (Jameson 1991: 412), "appropriating" individuals to its circuits (Grossberg 1992: 132). Here the language of flows attests not only to the pervasiveness and plasticity of capital but to its ultimate freedom from the boundedness of Identity. Capitalism becomes the everything everywhere of contemporary cultural representation.

If this catalogue seems concocted from exaggerations and omissions, that will not surprise us.[19] For we have devised it in line with our purposes, and have left out all manner of counter and alternative representations. Indeed, as our critics sometimes charge, we have constructed a "straw man" – or more accurately a bizarre and monstrous being that

[18] Arguments that capitalism is in fact "capitalisms" (see for example Pred and Watts 1992) may actually represent capitalism's chameleon qualities as an aspect of its sameness, its capacity for taking everything into itself. These arguments constitute capitalism as a powerful system that is not delineated by any particular economic practices or characteristics (except power). Everything in its vicinity is likely to be drawn into it, overpowered by it, subsumed to it. In related formulations, homogeneity, even of the economic kind, is not a requirement of a monolithic capitalism, since the nature of capitalism is "not to create an homogeneous social and economic system but rather to dominate and draw profit from the diversity and inequality that remain in permanence" (Berger 1980).

[19] In fact we were inspired to some extent by Foucault in *The Order of Things*, where "orders" or classifications are made to appear strange or ridiculous as part of a strategy of denaturalization.

will never be found in pure form in any other text.[20] The question then becomes, what to do with the monster? Should we refine it, cut it down to size, render it once again acceptable, unremarkable, invisibly visible? Should we resituate it among its alter and counter representations, hoping thereby to minimize or mask its presence in social and cultural thought? These are familiar strategies for dealing with something so gauche and ungainly, so clearly and crudely larger than life.

But of course there are alternative ways of disposing of the creature, perhaps more conducive to its permanent relegation. Might we not take advantage of its exaggerated and outlandish presence, and the obviousness that attends it? We can see – it has been placed before us – that a (ridiculous) monster is afoot. It has consequently become "obvious" that our usual strategy is not to banish or slay it, but rather to tame it: hedge it with qualifications, rive it with contradictions, discipline

[20] Of course this could be said of most representations. Many people have assured us that "nobody" thinks any more that capitalism is heroic, systemic, self-reproducing, lawful, structural, naturally powerful, or whatever it is we are adducing. We have come to identify this "nobody" with the one invoked by Yogi Berra ("Nobody goes there any more. It's too crowded.").

We are reminded of the early 1970s when many people found feminist arguments about the existence of a regime of sexism or male dominance to be paranoid or hyperimaginative. Women often argued, for example, that the men they knew were not "like that" or that particular texts, events or relationships did not display the contours of such a regime. These individuals were quite right to note that what feminists described as male dominance was not ubiquitous or pervasive, and was not fully manifest in the behavior of individual men (as indeed feminist activists were often tempted to adduce), yet that did not mean there were no practices and conditions of male dominance. What it meant was that those practices and conditions were often subtle rather than blatant, slippery rather than firm, invisible as well as visible, or visible only from particular locations. It was no simple matter to "reveal" their existence, tangled as they were with their opposites, their disconfirmations and misrecognitions, their negations, their contradictory effects, their failures, their alternative interpretations, the resistances they called forth, the always different contexts that produced the specificity of their forms of existence.

Perhaps a better way of saying this is that feminists were required to produce a theoretical object (sexism or male dominance or patriarchy or the binary hierarchy of gender) and to constitute it as an object of popular discourse and political struggle. That object was no more self-evident than any other (than, for example, the existence of something called "capitalism" before Marx did his work). In this sense, the burden can be seen to lie with us, to produce the discursive object of our critique. Those who invoke the "straw man" argument are questioning the initiative of constituting this theoretical object (by arguing that our construct is illegitimate in comparison to some other) and calling upon a putative community of understanding (of the real or right way to represent capitalism) to regulate the production of social and economic theory. But they are also reacting against the exaggerated appearance of capitalism as it is portrayed here. Presumably their intention would be to mute and domesticate that appearance rather than to highlight it as an object of criticism and derision.

it with contingencies of politics or culture; make it more "realistic" and reasonable, more complex, less embarrassing, less outrageous. But where does such a process of domestication leave us?

Unfortunately, it does not necessarily address the discursive features and figurings that render capitalism superior to its noncapitalist others. Capitalism might still relate to noncapitalist economic sites (in the so-called Third World and in "backward" regions and sectors in the developed world) through images of penetration. Its body could continue to "cover" the space of the social, so that everything noncapitalist was also capitalist (not of course a reciprocal relation). It could still be inherently capable of initiating thoroughgoing (perhaps dysfunctional) social transformation, relegating noncapitalism to a space of necessary weakness and defeat. It might still be driven by internal dynamics of expansion or regeneration, taking advantage of the relative vitality and longevity such imperatives confer. And it could still figure as a systemic totality, producing economic monism as an implication or effect. It seems quite likely, then, that noncapitalism could continue to be suppressed or marginalized by a tamer beast.

In the hierarchical relation of capitalism to noncapitalism lies (entrapped) the possibility of theorizing economic difference, of supplanting the discourse of capitalist hegemony with a plurality and heterogeneity of economic forms. Liberating that possibility is an anti-essentialist project, and perhaps the principal aim of this book.[21] But it is no simple matter to know how to proceed. Casting about for a way to begin we have found feminist and other anti-essentialist projects of rethinking identity and social hegemony particularly fruitful.

Strategy 2: Deconstructing the capitalism/noncapitalism relation

In the writings of Ernesto Laclau and Chantal Mouffe (1985, for example) we find the identity of "the social" rethought and decentered. Society resists being thought as a natural unity (like an organism or body) or as one that is closed by a structure, like patriarchy or capitalism, around a central antagonism or fundamental relation. Rather society can be seen as transiently and partially unified by temporary fixings of meaning. These are achieved in part through political struggles that change the relationship of social elements one to another.

Often though not always, the elements of society are articulated,

[21] In other words, this is a project of attempting to make difference rather than sameness "obvious," in the way that Sedgwick does for sexuality (1990: 25–6).

"sutured" as moments in a "hegemonic" relational structure. But this articulation is always ever incomplete and temporary, susceptible to subversion by the "surplus of meaning" of its moments (each of which has various "identities" in the sense of being differentiated within alternative relational systems). Thus the term "woman" has a different meaning when it is articulated with "private life" and "marriage" than when it is set in the context of "feminism" and "lesbian," and the latter contextualization is destabilizing to concepts of male prerogative associated with the former.[22] Identity, whether of the subject or of society, cannot therefore be seen as the property of a bounded and centered being that reveals itself in history. Instead identity is open, incomplete, multiple, shifting. In the words of Mouffe (1995) and other poststructuralist theorists, identity is hybridized and nomadic.

Perhaps we may pursue this further, into a region that is somewhat less traveled, to consider what this might mean for the economy, to ask what a hybridized and nomadic "economic identity" might be. If Mouffe and Laclau have rethought the "social," translating what was formerly closed and singular to openness and multiplicity, what implications might such a rethinking have for the "economic"? It might suggest, at the very least, that the economy did not have to be thought as a bounded and unified space with a fixed capitalist identity. Perhaps the totality of the economic could be seen as a site of multiple forms of economy whose relations to each other are only ever partially fixed and always under subversion. It would be possible, then, to see contemporary discourses of capitalist hegemony as enacting a violence upon other forms of economy, requiring their subordination as a condition of capitalist dominance.[23]

In the frame of such a discursivist and pluralist vision, emerging feminist discourses of the noncapitalist household economy can be seen as potentially destabilizing to capitalism's hegemony.[24] By placing the term "capitalism" in a new relation to noncapitalist "household production," they make visible the discursive violence involved in theorizing household economic practices as "capitalist reproduction." The feminist intervention problematizes unitary or homogeneous notions of a capitalist

[22] We are indebted to Daly (1991: 91) for a version of this example.

[23] For a longer and more developed version of this argument, see Gibson-Graham (1995b).

[24] Feminist economics (as well as other branches of feminist social analysis) has focused attention on unpaid household labor and the production and distribution of use values in the household and on the relative absence of these in both mainstream and Marxist discourses of economy (Waring 1988; Beasley 1994). For studies of the household economy and household social relations, see Delphy and Leonard (1992), Folbre (1993), Fraad et al. (1994), among many others.

economy. It opens the question of the origins of economic monism and pushes us to consider what it might mean to call an economy "capitalist" when more hours of labor (over the life course of individuals) are spent in noncapitalist activity.[25] It is possible, then, that such an intervention could mark the inception of a new "hegemonic discourse" of economic difference and plurality.[26]

At the moment, however, the conditions of possibility of such a discourse are decidedly unpropitious. For both as a constituent and as an effect of capitalist hegemony, we encounter the general suppression and negation of economic difference; and in representations of noncapitalist forms of economy, we have found a set of subordinated and devalued states of being. What is generally visible in these representations is the insufficiency of noncapitalism with respect to capitalism rather than the positive role of noncapitalist economic practices in constituting a complex economy and determining capitalism's specific forms of existence.[27]

In encountering the subordination of noncapitalism, we confront a similar problem to that encountered by feminists attempting to reconceptualize binary gender. It is difficult if not impossible to posit binary difference that is not potentially subsumable to hierarchies of presence/absence, sufficiency/insufficiency, male/female, positivity/negation. Thus rather than constituting a diverse realm of heterogeneity and difference, representations of noncapitalism frequently become subsumed to the discourse of capitalist hegemony. To the extent that capitalism exists as a monolith and noncapitalism as an insufficiency or absence, the economy is not a plural space, a place of difference and struggle (for example, among capitalist and noncapitalist class identities). The question then presents itself, how do we get out of this capitalist place?

Here we may fruitfully turn to the work of those feminists who

[25] See Katz and Monk (1993). Of course there are many possible indicators (such as numbers of people working at any one time, or value of output) that could be used to suggest the relative size of the "household economy."

[26] This is just one example of the sort of problem and opportunity that arises when noncapitalist forms of economy are theorized as both existing in society and as suppressed in economic discourse.

[27] This should not be taken to mean that there are no theorists who pursue a "dialectical" conception of capitalism, examining the ways in which capitalist development is a condition of noncapitalist development, but that such approaches are not dominant or even prevalent. Certain postcolonial theorists (Sanyal 1995, for example) argue that capitalist development in the Third World involves the constitution and valorization of noncapitalist economic activities, which articulate with and participate in constituting capitalism itself.

have attempted to (re)theorize sexual difference, to escape – however temporarily and partially – from the terms of a binary hierarchy in which one term is deprived of positive being. For woman to be a set of specificities rather than the opposite, or complement, to Man, man must become a set of specificities as well. If Man is singular, if he is a self-identical and definite figure, then non-man becomes his negative, or functions as an indefinite and homogeneous ground against which Man's definite outlines may be seen. But if man himself is different from himself, then woman cannot be singularly defined as non-man. If there is no singular figure, there can be no singular other. The other becomes potentially specific, variously definite, an array of positivities rather than a negation or an amorphous ground. Thus the plural specificity of "men" is a condition of the positive existences and specificities of "women."[28]

By analogy here, the specificity of capitalism – its plural identity, if you like – becomes a condition of the existence of a discourse of noncapitalism as a set of positive and differentiated economic forms. Feudalisms, slaveries, independent forms of commodity production, non-market household economic relations and other types of economy may be seen as coexisting in a plural economic space – articulated with and overdetermining various capitalisms rather than necessarily subordinated or subsumed to a dominant self-identical being.

But in order for this to occur, capitalism must relate to itself as a difference rather than as a sameness or a replication. For if capitalism's identity is even partially immobile or fixed, if its inside is not fully constituted by its outside, if it is the site of an inevitability like the logics of profitability or accumulation, then it will necessarily be seen to operate as a constraint or a limit.[29] It becomes that to which other more mutable entities must adapt. (We see this today in both mainstream and left discussions of social and economic policy, where we are told that we may have democracy, or a pared-down welfare state, or prosperity, but only in the context of the [global capitalist] economy and what it will permit.) It is here that anti-essentialist strategies can begin to do their work. If there is no underlying commonality among capitalist instances, no essence of capitalism like expansionism or property ownership or power or

[28] Here we may see a feminist argument for anti-essentialist discourses of identity as a political strategy of discursive destabilization, drawn from the work of Irigaray (Daley 1994, Hazel 1994).

[29] This is the problem, for example, with theories of capitalist regulation that array their "models of development" on an invariant social skeleton centered on capital accumulation (see chapter 7), or with representations of capitalist enterprises as centered by an imperative of profitability (see chapter 8).

profitability or capital accumulation,[30] then capitalism must adapt to (be constituted by) other forms of economy just as they must adapt to (be constituted by) it. Theorizing capitalism itself as different from itself[31] – as having, in other words, no essential or coherent identity – multiplies (infinitely) the possibilities of alterity. At the same time, recontextualizing capitalism in a discourse of economic plurality destabilizes its presumptive hegemony. Hegemony becomes a feature not of capitalism itself but of a social articulation that is only temporarily fixed and always under subversion; and alternative economic discourses become the sites and instruments of struggles that may subvert capitalism's provisional and unstable dominance (if indeed such dominance is understood to exist).

Strategy 3: Overdetermination as an anti-essentialist practice

The capitalism whose hegemony is intrinsic never attains full concreteness. Its concrete manifestations, its local and historical contextualizations, are always only modifications or elaborations of a dominance that already (abstractly) exists.[32] When capitalism is unified by an abstract self-resemblance, a conceptual zone is liberated from contradiction. Each time the name of capitalism is invoked, a familiar figure is (re)imposed on the social landscape.

For capitalism to exist in difference – as a set of concrete specificities, or a category in self-contradiction – it becomes necessary to think the radical emptiness of every capitalist instance. Thus a capitalist site (a firm, industry, or economy) or a capitalist practice (exploitation of wage labor, distribution of surplus value) cannot appear as the concrete embodiment of an abstract capitalist essence. It has no invariant "inside" but is constituted by its continually changing and contradictory

[30] The similarity here to anti-essentialist reconceptualizations of "woman" should be apparent. As sexual dimorphism has increasingly become understood as a discursive construct, it has become more difficult to see gender as socially constructed and mutable in contrast to the supposedly immutable (because biologically given) category of sex. Thus, there is a tendency now to recognize as "women" those individuals who are temporarily identified by themselves and others as women (who are, in Althusser's terms [1971], interpellated by the ideology of binary gender) rather than to define the category in some invariant way. No commonality unifies all the instances of "woman" in this anti-essentialist formulation.

[31] This is a project which is arguably being undertaken by those working on capitalist embeddedness or "different capitalisms" (see, for example, Mitchell 1995 and chapter 8).

[32] Usually this dominance is guaranteed by a logic of profitability, a telos of expansion, an imperative of accumulation, a structure of ownership and control, or some other essential quality or feature.

"outsides."[33] In the words of Althusser, the "existing conditions" are its "conditions of existence" (1969: 208). In the terms that Althusser appropriated from psychoanalysis, reappropriated and reformulated by later Althusserians, a capitalist site or practice is "overdetermined": entirely (rather than residually) constituted by all other practices, processes, events.[34] The practice of social theory and analysis involves specifying and exploring some of these constitutive relations. This practice cannot build upon a secure epistemological foundation, or orient itself around an ontological given. It is itself a process of radical construction.

Through the theoretical lens of overdetermination, a capitalist site is an irreducible specificity. We may no more assume that a capitalist firm is interested in maximizing profits or exploitation than we may assume that an individual woman wants to bear and raise children, or that an American is interested in making money. When we refer to an economy-wide imperative of capital accumulation, we stand on the same unsafe ground (in the context of the anti-essentialist presumption of overdetermination) that we tread when we refer to a maternal instinct or a human drive to acquisition. If we define capitalist sites as involving the appropriation or distribution of surplus value, we cannot make any invariant associations between this process and particular structures of ownership, or distributions of power (or anything else), just as when we identify women by the wearing of dresses, we cannot draw any necessary conclusions about what's in the mind or under the skirt.

When Capitalism gives way to an array of capitalist differences, its noncapitalist other is released from singularity and subjection, becoming potentially visible as a differentiated multiplicity. And here the question

[33] This is the meaning of the concept of overdetermination elaborated by Resnick and Wolff (1987). As a theoretical starting place or ontological presumption, overdetermination involves an understanding of identities as continually and differentially constituted rather than as pre-existing their contexts or as having an invariant core. While it is quite common today to recognize "woman" as a term that lacks a stable referent, given the feminist and other work that has gone into producing anti-essentialist conceptions of personal identity, other kinds of identities – especially those that have a certain theoretical standing – may seem more justifiably construed as entailing sameness and invariance as a condition of intelligibility. We do not wish to deny that sameness is one of the conditions of meaning, but we would understand it more as an enabling belief (that we are talking about the same thing) than as an actual state of ontological or conceptual "commonality." Furthermore, we believe that it is as important for leftists to decenter and destabilize "capitalism" as it has been for many feminists to undermine the presumed commonalities of "women."

[34] If overdetermination appears to conflict with the requirement of categorical invariance, that is precisely its function as a positive practice of anti-essentialism (see footnote 33 and chapter 2).

becomes, how might we want to specify their positive (if not finally definitive) beings? For certainly economic space could be divided and differentiated in any number of ways, some of which may be already quite familiar.[35] In this book we have chosen to proliferate differences in the dimension of class, but this is only one potential matrix of differentiation.

Strategy 4: Elaborating a theory of economic difference

Drawing on the Marxian tradition, which they understand as encompassing an existing discourse of economic difference, Stephen Resnick and Richard Wolff distinguish a variety of economic processes including (1) the appropriation of surplus labor and (2) its distribution, which they identify as "class" processes (these are the exploitative and distributive processes that Marx explored in their capitalist forms in volumes I and III of *Capital*). When individuals labor beyond what is necessary for their own reproduction and the "surplus" fruits of their labor are appropriated by others (or themselves), and when that surplus is distributed to its social destinations, then we may recognize the processes of class.[36]

Class processes of exploitation and surplus distribution can be understood as potentially taking place in all sites where work is performed – households, family businesses, communal or collective enterprises, churches, schools, capitalist firms and all the other sites of economic activity that are generally subsumed under the umbrella of "capitalism."[37] But by differentiating and separating the various forms of class processes, we create the possibility of theorizing the interactions between them.

[35] Theorizing difference in processes of exchange, for example, we are at once confronted with the traditional distinction between commodity and noncommodity exchange; and the domain of commodity exchange is itself fractured by a variety of class relations, since commodities (goods and services transacted in a market) may be produced under familial, capitalist, independent, slave and other relations of production. Certainly, there is "nothing simple" about a commodity.

[36] In this anti-essentialist formulation the appropriation of surplus labor is not conflated with power or property relations in the definition of class (see Wolff and Resnick 1986; Resnick and Wolff 1987; and chapter 3).

[37] As an example of noncapitalist activity subsumed to capitalism, Watts describes contract farmers in Gambia as "nominally independent growers [who] retain the illusion of autonomy but have become in practice what Lenin labeled 'propertied proletarians', de facto workers cultivating company crops on private allotments" (Pred and Watts 1992: 82). While Lenin was interested in demonstrating the extent of capitalist penetration and proletarianization as an indication of revolutionary readiness, it is not clear why Watts would want to argue that these instances of contract farming should be seen as subsumed to, rather than as different from and articulated with, capitalist practices and institutions.

This move also undermines the presumptive or inherent dominance of capitalist class relations. When capitalism is represented as one among many forms of economy (characterized, say, by the presence of wage labor and the appropriation of surplus labor in value form), its hegemony must be theorized rather than presupposed. Economic sites that have usually been seen as homogeneously capitalist may be re-envisioned as sites of economic difference, where a variety of capitalist and noncapitalist class processes interact.

One example may convey some of the potential power of such a re-envisioning. In chapter 6, where we examine discourses of globalization, we briefly consider the international finance sector, which is often represented as the ultimate flowering of capitalism. Yet what can we say is necessarily capitalist about this industry, if we examine – with an eye to theorizing economic difference – its production relations, the sources of its revenues, and the destinations of its loans and investments? To the extent that firms in the finance sector are engaged in commodity production, some will be capitalist sites where surplus labor is appropriated as surplus value from employees whereas others will be sites of independent commodity production – for example, the personal investment manager who is a self-employed entrepreneur and appropriates her own surplus labor – and therefore noncapitalist. Other noncapitalist enterprises within the industry will be the sites of collective production and appropriation of surplus labor.[38] It is not clear what it means to call the industry capitalist given these differences in production relations, except that it entails obscuring rather than illuminating plurality and difference. Moreover the revenues that are accrued by the industry can be viewed as having entirely heterogeneous sources (some are distributions of surplus value in the form of interest payments from capitalist enterprises; some come from noncapitalist enterprises including independent producers, sites of enslavement and sites of collective or communal surplus appropriation; some are consumer interest payments, that is, nonclass revenues in the terms of Resnick and Wolff and therefore neither capitalist nor noncapitalist). Finally, the investment and lending activity undertaken by the industry can be seen as an unruly generative force that is not entirely disciplined by the imperative of capitalist reproduction.

Indeed, it is easy to tell a story that highlights the unprecedented opportunities this industry has created for the development of noncapitalist class relations: for instance, the huge increase in "consumer" credit has made it much easier for small businesses (including collectives and

[38] Partnerships, for example, in which the surplus – including profit – is jointly appropriated and decisions about its distribution are jointly made.

self-employed producers as well as small capitalist firms) to obtain needed inputs like equipment and supplies through credit card purchases. This growth in unmonitored business lending has undoubtedly contributed to the success and viability of a large number of noncapitalist enterprises, and especially to the growing practice of self-employment. Thus even if one theorizes the finance industry itself as thoroughly capitalist, it can be represented as existing in a process of self-contradiction rather than self-replication – in the sense that it is a condition of existence of noncapitalist as well as capitalist activities and relations. A frothy spawn of economic diversity slips out from under the voluminous skirts of the (demon capitalist) finance industry.

In the context of a capitalist monolith, where class is reduced to two fundamental class positions, sometimes supplemented by intermediate or ambiguous class locations, individuals are often seen as members of an objectively defined or subjectively identified social grouping that constitutes their "class." In the discursive space of diverse class processes, on the other hand, individuals may participate in a variety of class processes at one moment and over time. Their class identities are therefore potentially multiple and shifting.[39] Their class struggles (over exploitation, or over the distribution of its fruits) may be interpersonal and may not necessarily involve affiliation with a group.[40] What this means for a politics of class transformation is interesting but of course uncertain. It is clear, however, that a discourse of class exploitation and surplus distribution – and the theoretical vision of the variety of their forms – might enable some individuals to understand their economic experience as both a domain of difference and a region of possibility: the possibility, for example, of establishing communal or collective forms of appropriation, or becoming self-appropriating, or reducing the surplus that is appropriated by others, or changing the destination and size of surplus distributions.[41] How these possibilities might articulate with visions (and realities) of economic "improvement" or "liberation" or "equality" is an open question. The answers to this question are to be

39 For example, a person may appropriate surplus labor from a partner at home, produce surplus labor at a capitalist place of work, and both produce and appropriate surplus labor as a self-employed entrepreneur. None of these class positions confers a fixed or singular class identity. Within one individual multiple class identities will overdetermine and contradict one another, as well as other positions of the subject.

40 In chapter 3 we offer an extended discussion of class.

41 Here we might imagine new sorts of alliances between managers and unions, for example, in capitalist firms, who might have common interests in reducing distributions of surplus value to financiers and instituting an Employee Stock Ownership Plan or other arrangement through which distributions to both unionized and non-unionized employees would be increased (see chapter 8).

constructed not only in theory but also perhaps through an anticapitalist politics of economic innovation.[42]

Strategy 5: Making do with the wreckage and rudiments

This book is founded upon a desire for deliverance from a capitalist present and future that offers little possibility of escape. But to the extent that we gain a certain freedom through the thinking and writing of the book, we lose as a consequence the positive force of our desire. We may struggle and strain to banish a hideous monster from our economic space. But our attempts at banishment and evacuation leave us in an impoverished landscape, full of lackluster abstractions ("difference") and emaciated categories ("noncapitalist class processes"). Freedom from "capitalism" has perhaps become imaginable (freedom at least of a discursive sort). But we leave behind us a creature larger than life and twice as exciting, to enter into a starveling's embrace.

Nevertheless we have embarked, or opened the possibility of embarking, upon a project that has a discernible logic and momentum. That project is to produce economic knowledge within (and by developing) a discourse of economic difference, and specifically a discourse of class.[43] At the outset, class as a category seems mundane and uncompelling, shorn of the consequence and privilege it enjoyed as the principal axis of antagonism in a unified capitalist space. The different forms of class processes are merely part of an "economy" that encompasses innumerable other processes – exchange, speculation, waste, production, plunder, consumption, hoarding, innovation, competition, predation – none of which can be said (outside of a particular discursive or political context) to be less important or consequential than exploitation. Situating and specifying class (and differentiating the many noncapitalist forms of class relations) is a theoretical process that involves discursively constructing the connections and contradictions between class and other social processes and relations, over small or great spans of space and time. In this process, the emaciated class categories will take on flesh. As they become embedded in stories and contexts, their emptinesses will be filled, their skeletal outlines plumped up by their "constitutive outsides." They will gather meaning and visibility, import and inflection. Narratives and

[42] Of course the eradication of capitalism may not be the object of such political projects, once capitalism is dissociated from images of necessary rapacity and predation, and from related tendencies toward economic monism or hegemonism.

[43] In this latter effort we are not alone (see, for example, the journal *Rethinking Marxism*). See also Gibson-Graham et al. (1997) where we bring together writings on class, economic difference, and subjectivity.

social representations of existing and potential alternatives to capitalism may begin to resonate, to generate affect, to interpellate subjects, to ignite desire. In other words, they may become compelling, just as so many representations of capitalism now are.

Here at the outset, however, the Identity of "capitalism" is for us much more compelling than the non-identity of "different class processes." We are still attuned to social narratives and images in which capitalism constitutes a powerful and pervasive presence, one whose social and economic ramifications are largely malign. Such representations call forth intense feelings and interpellate us as revolutionary antagonists to a capitalist economic system. In the absence of a "capitalist system" and the narratives that constitute and attend it, we feel an absence of the political emotions that are traditionally associated with anticapitalist politics. In slaying the capitalist monster, we have eliminated as well the subject position of its opponent.

This suggests that we may need to produce a noncapitalist economic imaginary in the absence of desire (or in the presence of multiple and contradictory desires). Whereas we may "desire" the "capitalist totality" because of the powerful antagonistic sentiments we feel in its vicinity, we may not want to live with it. We may want instead a landscape of economic difference, in the presence of which paradoxically we feel no desire. The process of social representation calls forth and constitutes desiring subjects – persons with economic, professional, sexual, political, and innumerable other compulsions and desires. But the representation of noncapitalist class processes has barely begun. Developing an economic imaginary populated with "friendly monsters" of the noncapitalist sort is itself a project – only minimally engaged in this book but underway in other locations – that has the potential to create new political subjects and desires.

For now, in this book, we will take only a few initial and rudimentary steps. We must starve capitalism's bloated body and invigorate its "constitutive outside" – these are the conditions of both envisioning "different capitalisms" and constituting a positive space of noncapitalist economic difference. Through this project of undermining and construction, we may begin the process of engendering new political visions, projects and emotions. Luckily this is a project we do not undertake by ourselves.

Representations of capitalism as political culture: a road map

We have chosen to focus this book primarily upon representations of capitalism, which we see as a formidable obstacle to theorizing and envisioning economic (and specifically class) difference. In terms of the strategies set forth in this chapter, then, we have largely pursued strategies 1 through 3. These involve us in delineating the object of our critique (the hideous and hegemonic monster) and in undermining the representational coherences, correspondences and naturalizations that attend it.

So many and mutually reinforcing are the representations of capitalism, and so diverse are their origins and confluences, that we have sometimes felt quite daunted in the face of the capitalist eminence. Much as we now see economic development politics as taking on "the economy" in localized skirmishes, we have seen ourselves as taking on "capitalism" in brief bouts and fragmentary encounters. These small ways of contending with a large creature, linked together as the chapters of a book, may present both gaps and overlaps to a reader. We can only hope that she or he will experience the former as relief and the latter as needed reinforcement.

In a sense, the book starts with chapter 11, which began its life as a talk at a large conference on Marxism. Attempting to understand why there might be so much antagonism to capitalism, but at the same time so little politics focused on constructing noncapitalist alternatives, the chapter addresses the ways in which certain kinds of Marxian economic theory have become an obstacle rather than a spur to anticapitalist political projects. We see chapter 11 as a kind of companion to this first chapter, encapsulating the themes and import of the book. One way to read the book might be to read chapter 11 next.

Chapter 2 finds its companion in chapter 10, in the sense that they are both focused on methods of "deconstruction" and categorical destabilization. In the earlier chapter we explore the Althusserian concept and practice of overdetermination – its potential both for emptying the category "capitalism" and for filling it up differently. Chapter 10 finds in Derrida's recent book on Marx certain instabilities in the category "capitalism" that represent traces of or openings for noncapitalism in the present and proximate future.

Chapter 3 introduces "class" in its anti-essentialist conceptualization, suggesting a range of noncapitalist class relations on the contemporary economic scene. But we must look to chapter 9 for a fully developed exploration of a noncapitalist class process and its interactions with a capitalist one.

In chapter 8, which is also an offspring of chapter 3, we consider distributive class processes and explore capitalism itself as a difference. This chapter represents the capitalist enterprise as a decentered and differentiated site, where the process of exploitation (the production and appropriation of surplus value) can be seen as producing a "condensation" of wealth. Focusing on the enterprise as a collection point from which wealth is dispersed in any number of directions, it suggests some of the contours of a new class politics of distribution.

In chapter 4 we explore both metaphorical and social space as colonies of capitalism and the phallus, where all objects are located and identified with respect to these master terms. Inspired by feminist representations of space and the body, we attempt to imagine spaces of becoming and difference, perhaps harboring or generative of noncapitalist forms. These themes are taken up in chapter 6 on globalization, where we attempt to undermine the "rape script" that structures globalization stories as narratives of capitalist penetration and dissemination.

In chapter 5 we interrogate the body metaphors that inform economic policy discourse, recognizing in systemic and organicist conceptions some of the origins of economic monism. In addition, we examine the ladder of evolution that sets economic development upon a single path (with capitalist development as its pinnacle). Drawing upon feminist rethinkings of the body and upon nonlinear conceptions of biological evolution, we attempt to undermine the notion of a unitary and centered (capitalist) economy pursuing a unidirectional development trajectory.

Following and extending the arguments of chapter 5, chapter 7 takes on the discourses of Fordism and post-Fordism, scrutinizing not only the conceptions of economic totality they embody but also the economic activism they have engendered. In both theory and practice, these discourses can be seen to be conditions of capitalist reproduction.

Each of these chapters represents a skirmish with the capitalist beast. In every encounter we depict the object of our obsession as powerful and well developed, but we also try to muzzle and silence it. Rather than giving it a platform from which to speak its dominance, as leftists including ourselves have often done, we enshroud it in a productive silence, in order that glimmers and murmurings of noncapitalism might be seen or heard. Perhaps these glimpses and low sounds will be tantalizing (or frustrating) enough to inspire some others to pursue them.

2

Capitalism and Anti-essentialism: An Encounter in Contradiction

Contemporary social theory is arguably the site of a dominant anti-essentialism, particularly in projects that seek to address philosophical and personal "identity."[1] To varying degrees the essentialist hold of biological, psychological and social (to name but three) universals and structures on the construction of identity has been loosened under the influence of poststructuralist critiques and reformulations. Rarely, however, is anti-essentialism associated with Marxian political economy, where the "economic" has often been privileged as the fundamental, necessary or *essential* constituent of social systems and historical events. Within political economy and the political movements it has spawned or inspired, economic determinism has reigned as the salient (though frequently criticized) form of essentialism, contributing to theoretical and political difficulties and even disasters.[2]

It is not surprising then that a definition of essentialism emanating from

[1] According to Resnick and Wolff, "*essentialism* mean(s) a specific presumption . . . that any apparent complexity – a person, a relationship, an historical occurrence, and so forth – can be analyzed to reveal a simplicity lying at its core" (1987: 2–3). Fuss defines essentialism as "a belief in true essence – that which is most irreducible, unchanging, and therefore constitutive of a given person or thing" (1989: 2). She argues that so obsessed have we become with overthrowing the strictures of essentialist thought in favor of social constructionism that anti-essentialism has itself become an "essence" of contemporary social theory.

[2] This does not mean that it has been unproductive, or unlinked to various kinds of success. But one has only to think of past revolutionary projects of "socialist" construction for an example of the negative impacts of an essentialist belief in the socially transformative effects of changes in production technology and the ownership of economic assets.

an anti-essentialist Marxism would stress the question of causality:

> In relation to conceptualizing causality, essentialism is the presumption that among the influences apparently producing any outcome, some can be shown to be *inessential* to its occurrence while others will be shown to be *essential* causes. Amid the multifaceted complexity of influences apparently surrounding, say, some historical event . . . , one or a subset of these influences is presumed to be the essential cause of the event. The goal of analysis for such an essentialist theory is then to find and express this essential cause and its mechanism of producing what is theorized as its effect. (Resnick and Wolff 1987: 3)

Within Marxism, anti-essentialist strategies that allow a rethinking of both identity *and* social determination will have a special resonance and purchase. It is for this reason that the Althusserian concept of "overdetermination" has been such a powerful contribution to Marxian anti-essentialism[3] and why, in discussing the contradictory encounter between Marxian political economy and anti-essentialist thought, I wish to start with Althusser.[4]

Althusser's legacy

Since the exciting days of the 1960s and 1970s when Althusser's influence upon English-speaking Marxists was comparable in scope and intensity to that of Derrida within contemporary cultural, literary, and feminist thought, a cavernous silence has until recently surrounded Althusser's contribution to contemporary social theory. The summary dismissal during the 1980s of his work as "structuralist" was accompanied by embarrassed attempts to distance the left from the tragic circumstances of his mental illness and the murder of his wife.[5]

Ten years ago, a guest editorial for *Society and Space* on the future of urban studies announced that the "sterile grip of Althusserian Marxism [had] been broken" (Saunders and Williams 1986: 393). What was so

[3] See, for example, the pioneering work of Resnick and Wolff (1987) in which Althusser's concept of overdetermination is developed and applied.

[4] In this and several other chapters we speak in the first person singular, for no simple reason that we can discern, except that our joint authorial persona allows for this possibility as well as for the apparently less fictional "we."

[5] ". . . (even) those of us who have regarded Althusser as a major reference point . . . reacted collectively by splitting (him) into good and bad objects: Althusser the philosopher, and Althusser the madman and murderer. And we saw this split in terms of a succession. The good Althusser of the 1960s and 1970s was simply replaced by the madman of the 1980s" (Montag 1995: 52). In his moving essay Montag goes on to affirm that no matter how hard Althusser tried to secure his own self-destruction and excommunication, his inspiration has remained present and alive.

striking about this pronouncement, in addition to its intimation that the post-Marxist moment was already and unambiguously upon us, was its attribution of "sterility" to a tradition that had spawned an enormous and rampant lineage, within urban studies and within social analysis more broadly understood. For precisely at that moment, in the mid-1980s, prominent research programs and theoretical concepts derived from Althusser's interpretations of Marx were forcefully shaping the content and directions of social research. At the same time, criticisms directed at these projects and conceptions could also be seen as deriving from positions elaborated by Althusser – reflecting both the contradictions within and the contradictory readings of his work.

Yet what the guest editorialists *accurately* perceived was the fact that Althusser himself was seldom to be seen or acknowledged as a source of both dominant theoretical conceptions and the criticisms of these conceptions that were beginning to emerge. Although his influence was everywhere to be encountered, Althusser the theorist was almost totally in eclipse.

The moment of the early to mid-1980s that brought together the height of Althusser's influence and the nadir of his reputation is now visible from a distance.[6] Recently and increasingly, Althusser's work has been acknowledged and it has become possible once again to situate oneself in his lineage (see for example Amariglio 1987; Resnick and Wolff 1987, 1989; Grosz 1990; Balibar 1991; Derrida 1993; Kaplan and Sprinker 1993; Callari and Ruccio 1996). Given the anti-essentialism animating much of contemporary social thought, it is not surprising to encounter the re-invocation of Althusser's voice. A central project of his work was to develop and elaborate certain anti-reductionist theoretical moments within the Marxian tradition, and one of the outcomes of that work has been to provide the theoretical and epistemological conditions for anti-essentialist Marxian political economy.

Reworking and extending the dialectic as he found it in Marx, Althusser borrowed from psychoanalysis the concept of overdetermination to mark his novel appropriation of dialectical thinking:

> (the dialectic) includes in the positive comprehension of the existing state of things at the same time also the comprehension of the negation of that state, of its inevitable breaking up; because it regards every developed form as in fluid movement and thus takes into account its transient nature, lets

[6] And its contradictory figuring of Althusser has become more apparent. It is now possible to see the ways in which Althusser's work has radically affected a diverse range of social theories, even ones that might generally be seen as diametrically opposed. We can trace, for example, his influences upon structuralist *and* poststructuralist feminisms and upon contemporary Marxisms *and* post-Marxisms.

nothing impose upon it, and is in its essence critical and revolutionary. (Althusser 1972: 175, quoting Marx)

Althusser's overdetermination can be variously (though not exhaustively) understood as signaling the irreducible specificity of every determination; the essential complexity – as opposed to the root simplicity – of every form of existence;[7] the openness or incompleteness of every identity; the ultimate unfixity of every meaning; and the correlate possibility of conceiving an acentric – Althusser uses the term "decentered"[8] – social totality that is not structured by the primacy of any social element or location.

The novelty, the originality, indeed, the virtual impossibility of thinking "overdetermination" in the face of classical philosophy and the western Enlightenment tradition is now widely acknowledged. In company with Derrida, Foucault, Deleuze/Guattari and following Spinoza, Nietzsche, Freud, and Wittgenstein – to name but a few of his contemporaries and antecedents – Althusser can be seen as contributing to a tradition (or more accurately a counter-tradition) that is intent upon undermining the certainties of western thought. Breaking with the conception of being, existence, or identity as entailing sameness and self-resemblance; breaking with the vision of real entities, concepts, and knowing subjects existing in relations of analogy, reflection, repetition or representation; breaking with the notion of identity as constituted by a definitive exclusion (of what it is not) that inevitably privileges some attributes as central while demoting and devaluing others; breaking with the practice of viewing historical events or social formations as constituted by certain processes more prominently than, or to the exclusion of, others;

7 Here we encounter a first contradiction, for of course the tendency to make overdetermination into an ontology (specifically a belief in the complexity and contradictoriness of being and identity) is itself to ascribe an essence to being and identity. Yet a very powerful anti-essentialism is unleashed and enabled by this ontological reading, since it suggests that any particular social analysis will never find the ultimate causes of events, for example, nor be able to definitively exclude the effectivity of any social or natural process in thinking the constitution of anything. The question about any relationship – between, say, industrialization and heterosexuality – becomes, not did one have anything to do with the other, or how important was one in the constitution of the other, but how do we wish to think the complex interaction between these two complexities?

8 In the glossary that accompanies *For Marx* and *Reading Capital* the term "decentred structure" (structure decentrée) is defined as follows: "The Hegelian totality presupposes an original, primary essence that lies behind the complex appearance that it has produced by externalization in history; hence it is a structure with a centre. The Marxist totality, however, is never separable in this way from the elements that constitute it, as each is the condition of existence of all the others; hence it has no centre . . . it is a decentred structure" (Althusser 1969: 253–4).

Althusser moved to displace the ontological and epistemological foundations of western thinking – destabilizing the oppositional hierarchies of positivity/negation, necessity/contingency, importance/inconsequence, reality/representation.

Via the (onto)logic of overdetermination, every identity is reconceived as uncentered, as in process and transition, as having no essence to which it will tend or revert. Every event is constituted by all the conditions existing in that moment (including the past and the future) or, in the words of Althusser, the "existing conditions" are its "conditions of existence" (1969: 208). No entity or occurrence can be said to exist as a clear boundedness exclusive of its exteriors. The process of existence implicates all exteriors, and by virtue of this implication undermines the hierarchy of importance that defines some attributes or causes as necessary or essential, and others as contingent or peripheral, to a particular locus of being. Here we may see Althusser's militance against the Idea, but also against the conventional dialectic of A/not A in which positive being entails its immanent negation; structured by a sameness and by a principal fracture, A/not A offers an intimation of becoming/supersession rather than opening each instance to multidimensional or complex contradiction (Resnick and Wolff 1989).

Althusser's concept of overdetermination can be seen as the site of a longing or desire – to resuscitate the suppressed, to make room for the absent, to see what is invisible, to account for what is unaccounted for, to experience what is forbidden. For Marxist political economy the hidden and unaccounted for has become – in the twentieth century and especially its political expressions – the effectivity of the noneconomic. In a potent sense Althusser's anti-reductionist project attempted to liberate the complex diversity of noneconomic social processes – not in a simple negation (or an inversion of emphasis and privilege) but within a plethora of effectivities of which the economic is one. One thinks, for example, of his powerful description of the "ruptural unity" that was the Russian revolution, occurring at the confluence of a vast heterogeneity of circumstances, currents, and contradictions (local, international, class and nonclass) in which "we can no longer talk of the sole, unique power of the general 'contradiction'" (referring to the economic contradiction between the productive forces and relations) (1969: 100):

> The "contradiction" is inseparable from the total structure of the social body in which it is found, inseparable from its formal *conditions* of existence, and even from the *instances* it governs; it is radically affected by them, determining, but also determined in one and the same movement, and determined by the various *levels* and *instances* of the social formation

it animates; it might be called *overdetermined in its principle*. (1969: 101)[9]

It is perhaps not too much to say that the concept of overdetermination was a key moment in generating an anti-economistic and anti-essentialist body of Marxian political economy and social analysis in the years since the publication of Althusser's early writings.[10] Here one thinks, for example, of the French regulation school(s) of political economy, of certain strands of socialist feminism, of the post-Marxism of Laclau and Mouffe, of critiques and reconceptualizations of economic development theory, of the postmodern orientation of the journal *Rethinking Marxism*. All of these traditions – some of which draw explicitly upon the theoretical strategy of overdetermination – have sought to undo the economic essentialism of key areas of Marxian theory. With respect to our particular aims in this book, it is interesting to consider the possibilities that these anti-essentialist projects present (sometimes by omission) for representing capitalism through the lens of overdetermination.

Althusser and capitalist regulation

Perhaps the most well-known research program to be constructed on the foundation (and via the "surpassing") of Althusser's work is the multivocal body of political economic thought known as "regulation" theory, from which come the concepts of Fordism and post-Fordism that have gained such currency on the left (see chapter 7). According to Alain Lipietz, who has very specifically delineated the regulationists' debt to and differences with Althusser (Lipietz 1993), Althusser enabled a vision of a complexly structured social totality made up of relations irreducibly multiple and various, without a center or origin, existing as

9 In the summary terms of a post-Althusserian conception of overdetermination, every entity or event exists at the nexus of a bewildering complexity of natural and social processes, constituting it as a site of contradiction, tension, difference, and instability (Resnick and Wolff 1987). Each overdetermined site or process participates in constituting all others; every cause is an effect, every relationship is a process of (inter)change and mutuality. The analysis of such complexities requires the adoption of an "entry point" that betrays the concerns of the analyst but cannot secure ontological priority or privilege.

10 I use the term anti-essentialist (rather than nonessentialist) to signal the impossibility of fully transcending essentialism, or even of wishing to. Anti-essentialism is a motive rather than an achievement, and even as a motive it cannot exist as a universal value or unmitigated good (so that though I identify myself as an anti-essentialist Marxist to distinguish my position within the Marxian tradition, I will on occasion intentionally or unavoidably practice various kinds of essentialism and will often respect or emulate it in the practice of others.) See Fuss (1989) for an extended discussion of this issue and a commitment to not essentializing anti-essentialism.

"a fabric, an articulation of relatively autonomous and specific relations, overdetermining one another" (p. 127). On the basis of this Althusserian social conception, the regulationists were able to theorize the forms and activities of the state, the institutions of civil society, and the realm of ideas and culture as something other than "supports for capital" (p. 112), and thus to conceive of the project of concretely specifying, for particular historical periods, how they might nevertheless come to play that role.

The regulationists' goal is to represent capitalist development as an overdetermined process in which capitalism is constituted by its social context rather than by forces internal to itself. They wish to depict capitalism as capitalisms, as a multiplicity, a set of specificities, rather than as a unity or a sameness. For this reason (among others) they replace the transhistorical concept of the mode of production with what Lipietz calls a "model of (capitalist) development," a structure that emerges contingently in particular times and places. The model of development is centered on a regime of capital accumulation (involving a macroeconomic balance between production and consumption), which is associated with a dominant technological paradigm and stabilized over a period of time by a mode of regulation. This regulatory complex – incorporating social norms, practices and institutions – "mediates" or "stalls" the contradictions that permeate capital accumulation and capitalist social relations. The formation of a mode of regulation permits capital accumulation to function smoothly and gives rise to extended periods of economic stability and capitalist growth.[11]

In Lipietz's view, models of development are the product of historical "accident" rather than the functional requirements of capitalist reproduction. Capitalism thus appears to be a contingent phenomenon, overdetermined by particular norms and institutions, constituted as a unique form of economy in a particular time and place. Yet the presence of logics – however much they may appear to be contextualized or under

[11] The models of development known as Fordism and post-Fordism have considerable currency both inside and outside the academy as ways of understanding the history of the "advanced capitalist countries" over the period since World War II. Fordism was characterized by a regime of accumulation in which mass production was coupled with mass consumption for approximately 25 years. This fortunate coupling was sustained by a mode of regulation that included the welfare state, the nuclear family, industrial unions and other institutions that promoted labor peace and consumer demand. When Fordism broke down (under the pressure of institutional rigidities, social demands, and limits to the industrial paradigm that had become exacerbated over time) the advanced capitalist nations entered a lengthy crisis period at the end of which post-Fordism emerged. Elements of the post-Fordist model of development include flexible production, life-style consumption, and the privatization of the functions of the welfare state (see chapter 7 for fuller characterizations of Fordism and post-Fordism.)

subversion – functions to isolate the economy from the rest of the social terrain.[12] Although capital accumulation is said to take place on a sustained basis under circumstances that only accidentally arise (and thus the definitive logic of capitalism appears to be entirely contingent upon external conditions) the logic of accumulation actually dominates the terrain of economic and social eventuation. It produces either instability and depression or, alternatively, growth and stability (the two states of being of the regulationist economy). The accidental development of the mode of regulation operates to change the manifestation of the logic of accumulation but not to displace it from the center of effectivity.

The model of development divides the field of contingency into two types of relation: (1) contingent relations that do not stabilize accumulation but have no effectivity and (2) relations that have effectivity but only as supports for accumulation. Only the latter are included within the compass of regulation theory. (Thus household relations can either be seen as contributing to stabilizing the regime of capital accumulation or can remain invisible within the space of theory.) The realm of accident is thereby deprived of its ability to contradict the logic of accumulation, and contingency is effectively subsumed to necessity.[13]

By virtue of their integration into a model of development that is centered on capital accumulation and fully coextensive with the social space, all social elements – or all that are visible within the space of theory – are marked as "belonging" to capitalism. The household appliances that transformed the conditions of housework in the Fordist era are visible, for example, primarily as consumption goods generated by capitalist mass production. Their alternative identity as productive technology in a noncapitalist setting is suppressed and devalued (even meaningless in the context of a "capitalist" model of development.) The norms and institutions that make up the mode of regulation are fixed in a singular system of meanings that is centered upon capital accumulation; they obtain their identities within a relational structure in which they

[12] The special rules that pertain within social theory for "the economy" are interesting to consider. Whereas "logics" of culture, or politics, or society seem quite theoretically unpalatable – the suggestion, for example, that universities or local governments are unified by a dominant logic or motive force seems an untenable reduction – "logics" in the economic realm are relatively acceptable and unremarkable. Thus enterprises and national economies are routinely viewed as subject to the requirements of a single principle such as profitability or accumulation. Similarly, while literary, cultural and social theorists are very willing to discuss the "textuality" of texts, landscapes and social worlds, there is considerable resistance to the idea that the economy can be treated as a text.

[13] See DeMartino (1992) and Laclau and Mouffe (1985) on the impossibility of positing contingency in a realm that is also inhabited by necessity and the related instability of all dualistic forms of determination.

function as capitalism's "supports."[14] Now there is no "exterior" that can subvert capitalism and define society (or economy) as also and always a noncapitalist space.[15]

Despite the rich and open-ended way in which the development of capitalism is depicted, the capitalism of the regulationists shows a familiar face. At the level of the economy as a whole, the logic of capital accumulation defines capitalism as a dynamic and growth-oriented economic system; at the enterprise level, the logics of profitability and competition promote technological dynamism and discipline individual capitals to adopt advanced production and marketing techniques. What is important for the argument at hand is that these economic logics unify all forms of capitalism as resemblances or replications. They thus constitute capitalism's immutable core and locate it within what Althusser would call an "empiricist" epistemology: the truth of the economic object is given in the object itself, and is available to the (universal) intelligence of the knowing subject. These invariant logics are the "essence" of capitalism. To the extent that they play this role, capitalism's identity is not relationally constructed within a specific social articulation, or through the contextualization of the economic term in a particular discursive setting, or through the situatedness and specificity

[14] Or else they are excluded from the picture. Though capitalism is not understood as determining the conditions that function to sustain it, only those conditions that sustain it are allowed within the theoretical space. (This is what Lipietz calls "ex post reproductionism" or "functionalism after the fact" [1993: 129]).

[15] It is ironic that regulation theory – and here I speak of only some of its forms – should produce a very fertile and anti-essentialist political economy and at the same time a conception of society ultimately structured as a unified capitalist space. One could explain this by observing that the regulationists start with the presumption of a capitalist economy, and with a capitalist society defined by its economic "base" – and that therefore it is only to be expected that a homogeneous social vision should emerge as their theoretical product. But that would involve seeing discourse as a linear process with origins (e.g., premises or assumptions) and ends (e.g., findings or representations) rather than as a recursive process involving the continual making (or undermining) of all its elements and conditions; it would thereby obscure the specific practices by which the economy and society is not only presumptively constituted but also discursively "reproduced" as a singularly capitalist place. In the arena of regulation theory, a number of practices contribute to this reproduction. First, the capitalist economy is endowed with essential logics but no counter or divergent logics emanate from other social locations. The effectivity of non-capitalism is thereby understated or negated (except in its operation as a support for capitalism) (Ruccio 1989). Second, only those social processes that sustain capital accumulation are incorporated within the social representation. This constitutes society as a "capitalist" totality, which presumably then has a capitalist base. Thus, not only do the regulationists presume that the economy is essentially capitalist and that society is somehow susceptible to the same designation, but the practice of their discourse provides the conditions of both these hegemonizing presumptions.

of the knower. Capitalism as a concept "resists" the destabilizing force of overdetermination. Its relative obduracy allows it to impose its identity on the social field and gives primacy to its dynamics in the complex unfolding of historical eventuation.

Socialist feminism and Althusser

Like the regulationists, certain early socialist feminists[16] drew on Althusser's celebrated "ISAs" essay in order to theorize the ways in which capitalism depends for its very existence on its constitutive "outsides."[17] In Althusser's conception, the full complex of economic, political and ideological processes and institutions is required for the process of social reproduction. In the context of a decentered social formation, not even capitalist production is more important to reproduction than are other social processes and dimensions.

When subjectivity, the family, the state, and civil society were acknowledged as equally consequential for social reproduction as the economic "base," the space of feminist theory was liberated – it seemed – from its secondary and subservient status. Women's activities in the family and in the household, and to a lesser extent in the public and voluntary sectors, became the focus of socialist feminist explorations of the "sphere of reproduction."[18] To the extent, however, that women's reproductive activities were understood to involve primarily the daily and generational reproduction of the capitalist labor force, the autonomous and effective

[16] When I speak of socialist feminism, I am speaking of a varied and ongoing tradition that is nested within the broader field of Marxist feminism (Moi 1985: 91–3). More recently the term "materialist feminism" has been coined to denote a new tradition, emerging from the engagement between Marxist feminism and postmodernism, which attempts to keep alive the interest fostered by socialist feminism in the social nature of women's oppression (Hennessy 1993: 5).

[17] In their attempts to produce a nonmechanistic and also nonteleological understanding of capitalist reproduction, the regulationists were also deeply influenced by the essay in *Lenin and Philosophy* (1971) entitled "Ideology and Ideological State Apparatuses." Here can be found a source of their emphasis on the conditions of reproduction, including the sphere of circulation and the integration of relations between Departments I and II; here also are specified many of the social institutions (ideological state apparatuses [ISAs]) that made their way into the concept of the mode of regulation.

[18] Michèle Barrett (1981: 19–29) points out the ambiguities in the feminist use of the term "reproduction," which may refer (in Althusser's sense) to the reproduction of the conditions of capitalist production, to biological reproduction, or to the reproduction of the labor force; only some of these activities are particularly associated with women, and none uniquely so. The ambiguities of the term were also complicated by the metaphorical spatialization of the concept as a "sphere" which was then mapped onto actual social locations; both the metaphorical and the actual spaces were assumed to be preeminently the spaces of women.

sphere uniquely accorded to women was in fact defined by the requirements of capital. Despite its effectivity and ostensible independence, it "belonged" to capitalism as "capitalist" reproduction.[19]

Some socialist feminists challenged economic monism by theorizing an alternative economic "identity" to capitalism rather than focusing on the conditions of existence of capitalist production (this project was a version of what became known as "dual systems" theory). Christine Delphy, for example, put forward a notion of patriarchy as a social system founded on a "domestic mode of production."[20] While this theoretical move could be seen as an attempt to fight economism with economism – in other words, to affirm the independence of patriarchy from capitalism by providing the former with an economic "base" – at the same time it operated to decenter society from capitalism and thereby to reopen the question of the "identity" of the economy and of the totality of the social.

In theorizing the domestic mode of production, Delphy gave noncapitalism (of the economic sort) a definite identity and constituted "the economy" as a space of difference inhabited by at least two (and possibly more) economic forms. Her theoretical vision thus opened up the possibility of a decentered social totality in which capitalism coexisted with other forms of economy in relations of mutual constitution and overdetermination.

The domestic or patriarchal mode of production[21] has encountered a number of theoretical difficulties and objections. Delphy has been criticized for importing class into the household (once again Marxism is seen to be imperialistically claiming the domain of gender as a space of class) and for deriving gender oppression from relations of (household) exploitation. At the same time her formulation and similar ones have often foundered on the requirement that a mode

[19] This does not mean that socialist feminist theory has been a monolithic body of thought primarily interested in subsuming the specific activities and oppression of women to the functional requirements of capitalism. Quite the contrary. Under the theoretical direction of socialist feminism, the sphere of reproduction became visible as an important locus of "relatively autonomous" social practices and institutions, and the family and civil society – even, in fact, the sphere of capitalist production – came to be seen as domains where women's specific oppression was reproduced in tandem with but not in subordination to wage labor and other conditions of capitalist production and exploitation.

[20] See Delphy (1984) and Delphy and Leonard (1992). In a similar theoretical intervention, Folbre (1987) theorized a "patriarchal mode of production" as the site of noncapitalist production and exploitation that was "articulated" with the capitalist mode of production in industrialized social formations as well as in those of the so-called Third World.

[21] Folbre (1987) uses this latter term.

of production take the theoretical form associated with capitalism, for example, that it operate as a system with laws of motion and crisis tendencies (Connell 1987). This requirement positions capitalism as the universal form or model of economy to which other forms of economy must aspire.

Just as man is the universal subject and species standard of phallocentric discourse, capitalism is positioned as the economic standard in the discourses I have called "capitalocentric."[22] In the presence of phallocentrism sexual difference is implicitly negated, since human subjectivity takes a singular form; on the other hand, woman is constituted as less than human since she is other to man. In the analogous context of capitalocentrism, capitalism which is actually a specific economic form becomes the very model or definition of economy. By virtue of their differences from capitalism all other forms of economy fail to conform to true economic specifications. In a way that is entirely familiar but nevertheless theoretically quite intractable, difference is rendered as "absence" or lack rather than as autonomous being.

The problem of capitalocentrism also arose with respect to the relative positioning of the capitalist and patriarchal systems. The "dual systems" project did not reject the *idea* of the primacy of certain social aspects, or the *idea* of an origin of social organization: it nominated different (or additional) primates and origins. To the extent that this was the case, dual systems theory did not open the identity of the social to incompleteness and undecidability.[23] Instead it gave society an alternative (binary) closure. This closure implicitly subsumed, and subordinated the social dimensions of race, sexuality, and other differences or antagonisms, to two "essential" structures that themselves tended to be arranged in a hierarchy of power and effectivity. While it made women's labor visible and prominent, dual systems theory had difficulty rescuing this labor from its subordinate status. The binary structure of the dual society was constituted and captured by capitalocentrism.[24]

22 Noncapitalist forms of economy are positioned within "capitalocentric" discourses as the opposites, the subordinates and servants, the replications, or the deficient, nonexistent or even unimaginable others of capitalism. In this case, for example, noncapitalist forms of economy are understood to be structurally or functionally similar to capitalism. The definition of capitalocentrism is based on an analogy with phallocentric discourse in which woman is the same, the opposite, or the complement of man (Grosz 1990: 150).

23 Or, in Althusser's terms, overdetermination.

24 For what might be called a "new socialist feminism" developed from the perspective of an anti-essentialist Marxism, see Fraad et al. (1994) and Cameron (1995) who undertake overdeterminist readings of household relations of exploitation (see also chapter 9).

Post-Althusserian post-Marxism

Centered social representations such as those embodied in regulation theory and a certain type of socialist feminist discourse are the explicit targets of Laclau and Mouffe's *Hegemony and Socialist Strategy*. Pointing in the Marxist tradition to the univocality of structure, to the fixity of meaning of all the elements that is ultimately traceable to the economic instance, Laclau and Mouffe argue that a closed social structure centered on the economy ultimately restricts revolutionary politics to the politics of class. The singularity of the social order positions a single collectivity as its true antagonist and transformer; thus the identity of revolutionary subjects is fixed by the centered structure just as are all the other elements of the social order. Such an essentialist prespecification of political subjectivity is counterposed in their formulation to an open-ended process of social construction in which social meanings, structures and subject positions are always being fixed and always under subversion.[25]

In displacing the economy from its founding and unifying role, Laclau and Mouffe theorize the social (and by implication the economic) as constituted within a multiplicity of discourses, each of which represents an attempt to fix social meanings and positions. Each social element therefore has a surplus of meanings (and exists in a variety of discursive relations to other elements) rather than holding a fixed or preestablished social location. Society is not constituted as a closed totality in which each part has a stable relation to the others. Rather it is transiently and partially unified by temporary discursive fixings. The "exteriors" that each discourse suppresses will always subvert the stability of the discursive structuring. Thus society is divorced from "structure" in the familiar sense of a rigid and definitive skeleton that gives its elements a single identity and specified function. The totality of the social is instead to be understood via the concept of overdetermination which Laclau and Mouffe adopt from Althusser. Conceived as the participation of

[25] Derrida (1978) distinguishes as follows between centered and decentered conceptions of social structure: "The concept of centered structure is in fact the concept of a play based on a fundamental ground, a play constituted on the basis of a fundamental immobility and a reassuring certitude, which itself is beyond the reach of play" (p. 279). Following upon what he describes as a ruptural "event" it has become no longer possible to "conceive of structure on the basis of a full presence which is beyond play" (p. 279). "Henceforth it was necessary to begin thinking that there was no center, that the center could not be thought in the form of a present-being . . . This was the moment when language invaded the universal problematic, the moment when, in the absence of a center or origin, everything became discourse . . . that is to say, a system in which the central signified, the original or transcendental signified, is never absolutely present outside a system of differences. The absence of the transcendental signified extends the domain and the play of signification infinitely" (p. 280).

every social aspect in the constitution of the others (and not in terms of mutual causality between elements constituted outside relations of overdetermination), overdetermination entails

> a break with orthodox essentialism not through the logical disaggregation of its categories – with a resultant fixing of the identity of the disaggregated elements – but through the critique of every type of fixity, through an affirmation of the incomplete, open and politically negotiable character of every identity. This was the logic of overdetermination. For it, the sense of every identity is overdetermined inasmuch as all literality appears as constitutively subverted and exceeded; far from there being an essentialist totalization, or a no less essentialist separation among objects, the presence of some objects in others prevents any of their identities from being fixed. (p. 104)

The concept of overdetermination displaces the capitalist economy from its traditional role within Marxism of fixing the meanings of all social practices and institutions. Presumably, then, for Laclau and Mouffe the economy can become a space of temporary political fixings and ongoing subversions. Like the rest of society its identity can be constituted within a relational system that is only provisionally hegemonic. Not only is it defined by its discursive context, but its meaning is never singular or complete.

Given this highly developed theoretical position it is startling to find "capitalism" in the work of Laclau and Mouffe as the hero of a familiar narrative of social development, one in which it appears "as a unitary force" (Diskin and Sandler 1993: 40), totally and presumptively hegemonic, not only in the economic but the social domain:

> It is in the context of the reorganization which took place after the Second World War that a series of changes occurred at the level of social relations and a new hegemonic formation was consolidated . . . the decisive change is what Michel Aglietta has termed the transition from an extensive to an intensive regime of accumulation. The latter is characterized by the spread of capitalist relations of production to the whole set of social relations, and the subordination of the latter to the *logic of production for profit*. According to Aglietta the fundamental moment of this transition is the introduction of Fordism, which . . . is the articulation between a labour-process organized around the semi-automatic production line, and a mode of consumption characterized by the individual acquisition of commodities produced on a large scale for private consumption. This penetration of capitalist relations of production . . . was to transform society into a vast market in which new needs were ceaselessly created, and in which more and more of the products of human labour were turned into commodities. This "commodification" of social life destroyed previous social relations, replacing them with commodity relations through which

> the *logic of capitalist accumulation* penetrated into increasingly numerous spheres. Today it is not only as a seller of labour-power that the individual is subordinated to capital, but also through his or her incorporation into a multitude of other social relations: culture, free time, illness, education, sex and even death. There is practically no domain of individual or collective life which escapes capitalist relations. (Laclau and Mouffe 1985: 160–1, emphasis added)

This representation of capitalism in *Hegemony and Socialist Strategy* is notable not only for the extraordinary transformative capacity with which capitalism is endowed – it is given sole responsibility for a thoroughgoing historical transformation – but also for the familiarity and unremarkability of the depiction. In a diverse array of texts and traditions, capitalism is rendered as the "subject" of history, an agent that makes history but is not correspondingly "made." If it is affected and shaped by its social contexts, it is not equivalently "subjected." Instead it claims the terrain of the social as the arena of its self-realization.[26]

While undoing the closed and singular social totality, and unfixing society from its economic base, Laclau and Mouffe leave the economy theoretically untouched. It remains positive and homogeneous, inhabited by a set of logics that increasingly define the character of the social landscape (Diskin and Sandler 1993). As the inadvertent result of their theoretical silence, the economy has a fixed (if atheoretically specified) identity and capitalism itself has a fixed and transparent (or generic) meaning. Its definition and operations are independent of articulatory practices and discursive fixings; it can therefore be seen as "an abstraction with concrete effects" (in Laclau and Mouffe's wonderful critical phrase) rather than as a discursive moment that is relationally defined.

In the rendition of recent economic and social history quoted above, for example, capitalism inhabits the present as a concrete embodiment of its abstract description. Its internal imperatives of growth and expansion are manifest in history as its external form. No "exteriors" (discourses in which it has other meanings) operate to subvert its unity and self-resemblance. The immutable logics at the core of its being are independent of its social contexts (they always operate and are not fully

[26] Common to a variety of theoretical sites (including the ones discussed in this chapter) is a mood that might be called "capitalist triumphalism" (of the sort that Timpanaro [1970] identifies in Marx) which accompanies a detectable teleology. Capitalism has attained a state in which no external relations threaten to undermine it and in which internal relations are ultimately (if sometimes problematically) reproductive. As a fully revealed and realized being, capitalism has a completeness that is denied to other social practices and dimensions (indeed, in the case of Laclau and Mouffe, to society itself).

susceptible to being abridged). This gives capitalism, and by extension the economy, a disproportionate effectivity. Unlike other social practices and processes in *Hegemony and Socialist Strategy*, capitalism both has and is an essence. It is a cause without being to the same extent an effect. In this sense it exists outside overdetermination.[27]

Laclau and Mouffe's capitalism is the protagonist of a unified narrative of development that sets the political stage. But the capitalism they describe, and the heroic role they assign to it, is a remnant or borrowing (from other parts of the Marxian tradition) rather than a product of their own theoretical elaborations. For various reasons, including the "retreat" from economism inspired by Althusser, it is now the case that post-Marxist and cultural theorists often avoid constituting the economy as a theoretical object (perhaps theoretical avoidance is anti-economism's highest form.) By itself, this is not a fatal "omission" or a necessary source of theoretical deformations, since it would be impossible to problematize every social dimension and practice. But the "failure" to theorize the economy is inevitably associated with certain problematic effects. The language of social instances that divides society into economy, polity, and culture (or some other such partition) continues to function as the general conceptual frame within which particular social discourses are inscribed. Unless the economy is explicitly written out, or until it is deconstructively or positively rewritten, it will write itself into every text of social theory, in familiar and powerful ways. When it is not overtly theorized, it defines itself as capitalism because it lacks another name.[28]

Post-Althusserian development and postdevelopment theory

During the post-World War II period bourgeois economic development theory installed industrial capitalism at the pinnacle of economic and

[27] The characterization of economic processes as independent of overdetermination is taken from Diskin and Sandler (1993). This important article on *Hegemony and Socialist Strategy* scrutinizes the ways in which economic terms (both those that Laclau and Mouffe criticize and those they uncritically use) are assumed to carry their essentialist meanings with them.

[28] One could say, only partly facetiously, that "ignoring" the economy at the end of the 20th century is something like "ignoring" the deity at the beginning of the 15th. Even if he is not accorded explicit attention, even when he is deprived of theory and description, God is present as an origin, organizer, or ground. The absence of specification does not preclude him from his world but gives him instead the space and permission to work his effects unseen. In the case of capitalism, these effects are socially pervasive and thoroughly transformative: "there is practically no domain of individual or collective life which escapes capitalist relations" (Laclau and Mouffe 1985: 161).

social evolution, constituting the western experience of capitalist development as the model and measure for all the world. From a critical position within this imaginary, many leftists saw "development" as a capitalist phenomenon but one that left "developing" countries in a state of "underdevelopment" or economic and social deformation (e.g., Frank 1969; Wallerstein 1974). Both mainstream and left development discourses tended to see the process of capitalist development as initiating a wide-ranging process of social transformation, and differed primarily in how to read such a change: was it the coming of an earthly kingdom or, alternatively, was it largely a process of devastation and laying waste? Both theory and counter-theory tended to be economistic (in the sense that the economy was the principal motor of a thoroughgoing social transmutation) and monistic (in the sense that development was the creation of an integrated capitalist economy out of something heterogeneous and dispersed).

In the 1970s, on the basis of Althusser's work with Balibar (1970), a theoretical movement emerged on the left that challenged the economic monism of traditional development theory. This was the movement to theorize social formations as potentially involving the "articulation of (multiple) modes of production," especially in the Third World.[29] In Wolpe's (1980) conception, such an articulated heterogeneity does not necessarily presume capitalist dominance:

> the social formation is not given any necessary structure. It is conceived as a complex concrete object of investigation which may be structured by a single mode, or by a combination of modes none of which is dominant, or by a combination of modes one of which is dominant. (p. 34)

One of the goals of this theoretical project was to produce a discourse of economic difference, an alternative to representing the "complex conjuncture of the world economy" as an "undifferentiated" capitalist space (p. 30). But it also aimed to theorize noncapitalist forms of economy as having constitutive effects on capitalism itself (p. 4). Each of these features is distinctive (though not unprecedented) in the literature of Marxian political economy.

Interestingly, the problem of "capitalocentrism" became quite openly visible in the context of this Althusserian project, in part because individual theorists were in each case struggling with theorizing the relationship between capitalism and noncapitalism in a specific social site. Wolpe

[29] Wolpe's (1980) collection of essays entitled *The Articulation of Modes of Production* included work by Claude Meillassoux, Georges Dupré and Pierre-Philippe Rey, among others involved in this Althusserian theoretical movement.

notes, for example, that although there was no inherent functionalism in the concept of articulation of modes of production, in "the concrete analyses of social formations" (p. 41) theorists tended to represent the persistence of precapitalist modes of production as "functional for capital" (p. 40).[30]

Capitalocentrism in this context involves situating capitalism at the center of development narratives, thus tending to devalue or marginalize possibilities of noncapitalist development.[31] When independent commodity producers (like coffee growers in Central America) who own their own means of production, appropriate and distribute their own surplus labor, and buy and sell commodities on markets, are considered to be capitalist (the *same*), an opportunity to represent economic difference and to theorize the specificity of both capitalism and independent commodity production is lost. When noncapitalist forms of economy are coded as primitive, backward, stagnant, traditional, incapable of independent growth and development, and *opposed* to the modern, growth-oriented, and dynamic capitalist economy, development is defined as a process that necessitates the elimination or transformation of noncapitalist forms (or at least their subordination to capitalism). When noncapitalism is seen to articulate functionally (as the *complement*) to capitalism by providing, for example, underutilized savings and labor for capitalist industrialization, or seasonal labor power, or cheap means of production and consumption, its developmental effectivity is captured and contained by capitalist reproduction. When grassroots producer cooperatives or local development initiatives for self-employed producers are seen as existing *inside* a capitalist world economy, they become islands of noncapitalism in a capitalist sea, surviving and even thriving as isolated features in a landscape governed by the "laws" of capitalist development, or as "local" pockets of resistance and difference in a "global" system. Capitalism in all these cases is the central or dominant identity of "Third World," industrialized, and global

[30] Wolpe himself defines the notion of articulation in terms of the relationship between capitalism and precapitalism, identifying the latter not by its own characteristics but by its status as capitalism's precursor.

[31] In a contemporary project of theorizing economic difference, Sanyal (1995) offers a selfconsciously capitalocentric vision of postcolonial capitalism as a "dominant particular in a world of difference" which participates in generating the noncapitalist class processes that both sustain and undermine it. According to Sanyal the World Bank is now putting forward a new vision of developing economies as "overdetermined complexes of capital and noncapital," ironically producing a noncapitalocentric version of the postcolonialists' representation of class diversity and proliferation. Both of these representations foreground noncapitalism and give it a certain amount of autonomy. In these contexts the devaluation and marginalization of noncapitalist activities is relatively subtle.

economies. Recently, left theorists have moved to represent development as an economistic discourse rather than a universal social reality (see, for example, Escobar 1995). Development as a discourse is seen to have produced the "Third World" as a dependent identity subordinated to the management and surveillance of an international development bureaucracy. Contrary to promises and expectations, development has meant poverty and immiseration for "Third World" societies rather than inaugurating a "kingdom of abundance" (Escobar 1995: 4).

In the postdevelopment literature, anti-economism and anti-Eurocentrism work in tandem to displace monolithic images of the future and to bring into visibility cultural and social differences, resistances to hegemony, local power, dynamism and subjectivity. Just as the "articulation of modes of production" militated against the economic monism of traditional development theory, postdevelopment theory militates against the economic essentialism of both mainstream and left development theories and practices. Like the former project, however, postdevelopment theory shows a tendency toward "capitalocentrism":

> global capital . . . relies today not so much on homogenization of an exterior Third World as on its ability to consolidate diverse, heterogeneous social forms . . . Some of the peripheral forms take on [a] dissonant role because of their inadequacy in relation to their own national markets. This does not mean that they are less organized by capital . . . The minority social organizations of the tropical rain-forest areas, for instance, are not entirely coded or territorialized by capital (as are the formal urban economies). Yet to the extent that the economy constitutes a worldwide axiomatic, even these minor forms are the target of social subjections. The global economy must thus be understood as a decentered system with manifold apparatuses of capture – symbolic, economic, and political. It matters to investigate the particular ways in which each local group participates in this complex machinelike process, and how it can avoid the most exploitative mechanisms of capture of the capitalist megamachines. (Escobar 1995: 99)

Heterogeneities – including points of noncapitalism – may themselves be produced by capitalism:

> a political economy of global economic and cultural production must thus *explain* both the new forms of capital accumulation and the local discourses and practices through which the global forms are necessarily deployed; it must *explain* . . . "the production of cultural difference within a structured system of global political economy." (Escobar 1995: 98 emphasis added)[32]

[32] Quoting Pred and Watts (1992: 18).

In these representations the global (capitalist) economy is an origin of locally manifest diversities, which exist *inside* it both spatially and causally. The task of political economy is not to understand their existences differently but to produce specific versions of an already existing general understanding: to *explain* how global capitalism (which, in our reading, is the essence) gives rise to heterogeneity and diversity (the phenomena). This imperative (re)orients postdevelopment theory around the practice of an economic "science" which, ironically, is the target of its critiques of epistemological realism, economic essentialism and modernist developmentalism.

All this cannot of course detract from the novelty and importance of Escobar's work or from the magnitude of the postdevelopment project. It simply points to a way in which "development" exists *within* these projects, and a way in which capitalism exists outside overdetermination.

"Because the mountain grass cannot but keep the form where the mountain hare has lain"[33]

What interests me most here is the question of why the economism of which capitalism is the bearer is so difficult to moderate or excise. And what may account for the economic monism or hegemonism that accompanies most representations of capitalist society and development? Here a partial answer may be found in the metaphysics of identity that Althusser sought to undermine. Operating under an "imperative of unity" (Hazel 1994: 4) western conceptions of identity entail both the unity of an object with itself (its self-resemblance) and its one-to-one relation with the sign by which it is known: one word with one meaning, corresponding to one thing. To such an essentialist reading of identity "capitalism" designates an underlying commonality in the objects to which it refers. Thus we are not surprised to encounter a capitalism that is essentially the same in different times and places (despite the fact that sameness as the precondition of meaning is exactly what various structuralist and poststructuralist traditions have sought to undermine.) By virtue of their identification as capitalist settings, different societies become the sites of a resemblance or a replication. Complex processes of social development – commodification, industrialization, proletarianization, internationalization – become legible as the signatures of capitalism rather than as unique and decentered determinations.

When capitalism exists as a sameness, noncapitalism can only be subordinated or rendered invisible (like traditional or domestic economic

[33] W. B. Yeats, "Memory," lines 4–6.

forms). Noncapitalism is to capitalism as woman to man: an insufficiency until and unless it is released from the binary metaphysics of identity (where A is a unified self-identical being that excludes what it is not).[34] If capitalism/man can be understood as multiple and specific; if it is not a unity but a heterogeneity, not a sameness but a difference; if it is always becoming what it is not; if it incorporates difference within its decentered being; then noncapitalism/woman is released from its singular and subordinate status. There is no singularity of Form to constitute noncapitalism/woman as a simple negation or as the recessive ground against which the positive figure of capitalism/man is defined. To conceptualize capitalism/man as multiple and different is thus a condition of theorizing noncapitalism/woman as a set of specific, definite forms of being.

It is easy to appreciate the strategic effectiveness of reading the texts of capitalism deconstructively, discovering the surplus and contradictory meanings of the term, the places where capitalism is inhabited and constituted by noncapitalism, where it escapes the logic of sameness and is unable to maintain its ostensible self-identity (see chapter 10). But overdetermination can be used as an additional anti-essentialist theoretical strategy to complement and supplement the strategy of deconstruction. Taken together these strategies have the potential to undermine capitalism's discursive "hegemony" and to reconceptualize its role in social determination. Representations of society and economy cannot themselves be centered on a decentered and formless entity that is itself always different from itself, and that obtains its shifting and contradictory identity from the always changing exteriors that overdetermine it.

Just as postmodernism obtains its power from modernism (its power to undermine and destabilize, to oppose and contradict),[35] so can an overdeterminist approach realize its power and strategic capacity by virtue of its oppositional relation to the preeminent modes of understanding both language categories and identity/being. To the extent that we conceptualize entities as autonomous, bounded, and discrete (constituted by the exclusion of their outsides), and as the unique referents that give each sign a stable and singular meaning, to that extent does the strategy of thinking overdetermination have the power to destabilize theoretical discourse and reposition the concepts within

[34] Poststructuralist feminist theorists pursue a number of discursive strategies in their attempts to release woman from the negativity and "lack" of her position in the binary structure of gender identity. The efforts of Luce Irigaray, for example, to theorize sexual difference have been instructive in the context of our efforts to theorize economic difference.

[35] See Ruccio (1991).

it.[36] Through the lens of overdetermination, identities (like capitalism) can become visible as entirely constituted by their "external" conditions. With an overdeterminist strategy we may empty capitalism of its universal attributes and evacuate the essential and invariant logics that allow it to hegemonize the economic and social terrain. Overdetermination enables us to read the causality that is capitalism as coexisting with an infinity of other determinants, none of which can definitively be said to be less or more significant, while repositioning capitalism itself as an effect.

That the capitalist economy often escapes reconceptualization and so continues to function as an organizing moment, and an origin of meaning and causation in social theory, cannot be understood as a simple theoretical omission. It is also a reassertion of the hegemonic conceptions of language and determination that overdetermination is uniquely positioned to contradict. It is a testimony to the power of overdetermination that it has allowed certain post-Althusserian theorists to envision an "economy" that is not singular, centered, ordered or self-constituting, and that therefore is not capitalism's exclusive domain.[37] But it testifies to the resilience of the dominant conceptual context (it should perhaps be called a mode of thought) in which the objects of thought exist independently of thought and of each other that an autonomous economy still exists and operates in social representation.

One can say that representations of the capitalist economy as an independent entity informed by logics and exclusive of its exteriors have allowed capitalism to hegemonize both the economic and the social field. One can also say, however, that overdetermination is a discursive strategy that can potentially empty, fragment, decenter and open the economy, liberating discourses of economy and society from capitalism's embrace. But that process, far from being over or even well on its way, has hardly begun.

[36] For this reason, it doesn't make sense to think of overdetermination as an extra-discursive ontology which a discourse of overdetermination reflects. Overdetermination operates as a revolutionary discursive strategy only in the context of essentialist social and epistemological conceptions.

[37] See, for example, Cullenberg (1994), Fraad et al. (1994), McIntyre (1996).

3

Class and the Politics of "Identity"

Recent episodes of restructuring in industrialized societies seem to have been largely a negative experience for the "working class."[1] In the eyes of many observers, capital has achieved a new ascendancy, whether by virtue of its increased mobility and internationalization, or by virtue of a complex transition in society as a whole. By contrast, and by extension, workers have experienced declining standards of life and work, a decrease in bargaining leverage, and a general waning of effective militancy.

Given the new economic conditions that are widely acknowledged to characterize the 1980s and 1990s, many have given the restructured working class an unpromising political prognosis. The industries that were bastions of working-class militancy have declined, while at the same time we have seen the rise of high technology and service industries without the long tradition of solidarity and the unambiguous "working-class" image of traditional industrial jobs. The influx of women into the labor market and the increase in part-time and temporary jobs have created a labor force that is less likely to experience work as the primary basis of identity (Clark et al. 1986: 30). In general, then, it seems that in spite of (and perhaps even because of) the rapid proletarianization of women, both the work experience and the consciousness traditionally associated with the "working class" have declined, and with them the possibility of a viable politics of class.

Many recent studies of industrial change directly or indirectly communicate a discouraging picture of the potential for a contemporary class politics. In Los Angeles, for example, Soja (1989: 208) argues that industrial and urban restructuring have created a regional labor market

[1] Massey notes that "'industrial restructuring' is a process of class restructuring; it is one of the mechanisms by which the social structure is re-shaped, social relations changed and the basis for political action broken down or reconstructed" (1983: 74).

"more occupationally differentiated and socially segmented than ever before." The resultant demoralized and "K-marted" labor force experiences greater "social control than [has] hitherto marked the historical geography of capitalism" (Soja 1989: 207, 221). According to Storper and Scott (1989: 35) the successful development of "flexible production complexes" has occurred in "places without a prior history of Fordist industrialization, where the relations of production and work could be reconstructed anew." In many of these places, "neo-conservative attitudes about work and life have become remarkably pervasive." For Thrift (1987: 207), who sheds light on the dual processes of class formation and industrial change by examining the "new middle classes" emerging in late twentieth century capitalist societies, "the heroic age of class struggle has been replaced by a more prosaic age of class dealignment." In each of these cases, the decline of militant class politics is traced to the changing structure of industries and occupations, among other forces.

It would seem that capitalist industrial restructuring has broken down the "unity" of the working class, allowing differences to be played upon to the benefit of capital. Women are implicated in this process as the decline of traditional class politics has been accompanied by a realignment of gender and class. McDowell (1991: 417) argues that "the feminization of the labour market is amongst the most far-reaching of the changes of the last two decades." While men are increasingly subject to the terms of the feminized labor market, with its proliferation of part-time and temporary jobs, women have become a central component of the restructured labor force. Women and men constitute a "new" working class, one that has lost its industrial muscle:

> In the present era, it seems as if the interests of working class men and women are drawing closer together as both sexes are adversely affected by the reconstruction of large areas of work as "feminine." In this latest round in the continuous struggle over the control of women's labour, the majority of women *and* men are losing. Capital is the beneficiary. (McDowell 1991: 416)

At the moment, opportunities for working-class solidarity are overshadowed and perhaps jeopardized by the greatly increased demands on women both in the labor market and in the domestic realm, where social supports that were the hallmark of "patriarchal capitalism" have been withdrawn (McDowell 1991). For McDowell (1990) and Johnson (1990), the new sexual division of labor associated with restructured capitalist industry has intensified women's exploitation and reconstituted and reasserted gender oppression in the workplace and at home.

What we find interesting and alarming in the restructuring research program is that it has created, as one of its byproducts, a discourse of working-class decline and disempowerment. This discourse is often associated with intimations of the decreasing social importance and political relevance of class.[2] Narratives tracing the misfortunes of the traditional working class coalesce with images of economic development "beyond" class, producing a vision of the decline of class politics as a potent social force.

It is our view that images of class powerlessness, decline, and irrelevance have discursive as well as nondiscursive origins, and it is the former that we wish to address. In particular, we wish to focus on the role of particular conceptions of class in generating the disheartening prognoses for class politics that have emerged from the restructuring research program. In the discussion that follows, we first consider some of the ways in which class is understood in contemporary political economy and then present an alternative Marxian conception of class. In each case, we are attentive to the ways in which concepts of class are embedded in visions of society and implicated in conceptions of the political subject.

In the latter part of the chapter, we explore some ways in which an alternative theory of class and an associated re-theorization of social and personal "identity" can make visible a politics of class that is largely invisible in restructuring research. As one of the byproducts of our discussion, we suggest a new dimension to the articulation of gender and class. This is put forward not as a general resolution to the troubles that have plagued this relationship in theory, but as an example of the different political insights that may accompany a reconceptualization of class.

[2] Interestingly, these disclaimers about the relevance of class within the contemporary social context are not accompanied by any diminution in the number of references to class in the abstract (often in the company of race, gender and, more recently, sexuality) as a key axis of individual identity. It is almost as though, having been eradicated as a meaningful category of conjunctural social analysis, class makes its reappearance as an ontological given of "the social." In the current environment of discursive enforcement, it is difficult for social analysts to avoid affirming (by reference to the race-class-gender-sexuality nexus) a commitment to the ontological priority of these constituents of identity. The emptiness and unquestioned selectivity of this commitment is cause for reflection in this and subsequent chapters.

In remarking the peculiar absence of class in theories and analyses of contemporary subjectivity, we are not alone. Wendy Brown, among others, notes that the recitation of the "multiculturalist mantra, 'race, class, gender, sexuality'" (1995: 61) is generally associated with a silence about class. Nevertheless she detects some ways in which nonclass identity politics "are partly configured by a peculiarly shaped and peculiarly disguised form of class resentment" (p. 60), testifying to the effectivity of class despite its hidden and inarticulate position.

Class defined: problems of social and personal identity

Various concepts of class coexist within Marxian political economy, often within the writings of the same person. Without attempting an exhaustive survey, we can perhaps safely generalize that most Marxists understand the term "class" as referring primarily to a social group. Individuals are members of a class by virtue of some commonality, either structurally or experientially defined.

Three shared attributes and experiences are commonly invoked in defining social groups as classes. One of these is *power*, with control over the labor process and/or domination in other aspects of social life distinguishing ruling classes from the ruled. Classes may also be distinguished on the basis of *property ownership*, especially of the means of production. Finally, classes are defined by their relation to *exploitation*, the question of whether they produce surplus labor or appropriate it. All or several of these dimensions may be embraced in the term "relations of production," which is the most familiar marker of Marxian conceptions of class (Wolff and Resnick 1986).

Very often, two or all three of these dimensions are linked in a composite conception of class. Walker (1985: 169–70), for example, invokes what he calls the "classic triad – extraction of surplus, ownership of means of production, and control of the labor process" in characterizing Marx's "bare bones" definition of class. Massey (1984: 31), in a variation on this theme, sees two of the three dimensions entailing the third. She defines the bourgeoisie as having ownership and possession (the latter involving control of the labor process), while the working class is excluded from both. By virtue of this dual exclusion, the bourgeoisie is able to extract surplus value from the working class.

In general, there is dissatisfaction with a simple conception of society structured by two major classes that are constituted by the relations of production. While this conception is used as a kind of foundation, most social analysts go beyond it to embrace the complex multiplicity of classes and class locations in the historical setting of particular social formations. One of the most influential mappings of contemporary class structure is that developed by Wright (1978). Massey (1984) appropriates and modifies Wright's triangular representation of three major classes outlined by Marx – the bourgeoisie, the working class, and the petty bourgeoisie.[3] Along the axis between the bourgeoisie and the working class are many intermediate locations, distinguished

[3] Members of the petty bourgeoisie have both ownership and possession of the means of production which confers no control over the labor of others, since they have no employees.

by degrees of economic ownership and of possession, from managing director to production supervisor to laborer. Along the axis between working class and petty bourgeoisie are workers with varying degrees of control over the labor process, from self-employed to semi-autonomous worker to laborer. And along the axis between the bourgeoisie and the petty bourgeoisie are those with greater or less control over the labor of others. The proliferation of intermediate class locations refines and complicates the conception of class.[4]

While this type of elaboration is intended to rectify the over-simplifications of the two-class model, other formulations attempt to go beyond what is seen as the economism of theorizing class solely in terms of relations of production. Presuming that in order to be a class a group of individuals must achieve a unity other than a shared location in an economic structure, these formulations are concerned with "class formation" as a complex process involving political, cultural, ideological, and other forces (Wright 1985). Most often invoked are political processes, which may raise consciousness and help to transform a class-in-itself into a class-for-itself. Following Laclau (1977) and Thompson (1963) among others, many see classes as social groups constituted as an "effect of struggle," sometimes in the workplace but often in the arena of the community or the local or national state. And, increasingly, place is coming to be seen as an important constituent of actual classes (see, for example, Walker 1985; Thrift and Williams 1987; Massey 1984).

In general, these formulations create images of classes defined initially (or in the last instance) by the economy and comprised of individuals with "objective" (albeit sometimes ambiguous or contradictory) locations in the relations of production. But these objective conditions are understood as defining class only in the narrowest sense. A full or complex conception of class takes into account the ways in which groups are formed and the subjective bases of group identification:

> production relations indicate the sites of class relations in the economic structure, but those sites do not designate whole classes as integral, empirical groups of men and women. The fact that people occupy similar

4 In his subsequent work (for example, *Classes* [1985]) Wright dropped this framework for theorizing "contradictory class locations" in part because it emphasized the role of domination rather than exploitation in defining classes. Building on the work of Roemer (1982) Wright has been concerned to specify class locations on the basis of different kinds of exploitation arising from inequalities in the distribution of productive assets (i.e., property relations) (1985: 71–2). Individuals can occupy contradictory locations in which they are *exploited*, because they are excluded from ownership of the means of production, and yet are "opposed to workers because of their effective control of organization and skill assets" (1985: 87).

> places in the relations of production does not in itself imply any other empirical level of coherence, still less any kind of necessary political unity about pre-given common interests. Wright talks of class capacities, the social relations *within* a class which determine how internally coherent it is. All of which means . . . that "whole classes" are rarely actual political subjects . . . (Massey 1984: 43)

From our perspective, these interesting and complex conceptions of class contribute to a number of theoretical and political problems, including difficulties in conceptualizing class transformation and in theorizing individual and group identity in relation to class. If we define class in terms of power over the labor process, ownership of (industrial) property, and exploitation, how do we understand a situation in which one of these dimensions changes? When a capitalist industry is nationalized, for example, citizens and workers become owners of means of production. But what are the implications of this change in ownership for the production and appropriation of surplus labor in value form (that is, capitalist exploitation) and for worker control of the labor process? Neither democratic control of the workplace nor the end of capitalist exploitation necessarily or even readily follows from the nationalization of a particular industry or of an entire industrial system. A change in ownership, even a radical one, may not mean a transformation in other dimensions of class. Such a change confronts the analyst with a choice between theorizing an ambiguous instance (neither wholly capitalist nor wholly socialist, for example) or giving one aspect of the composite conception priority in the definition of class. Either choice may have profound political consequences. In the case of the former Soviet Union, for example, those who refused to emphasize ownership tended to see socialism as something that was yet to be achieved and the existing regime as something to mobilize against. Those, on the other hand, who saw ownership as the principal dimension of class were more likely to support and defend the existing regime (Resnick and Wolff 1994).

Similar problems arise when we attempt to understand individual class positions and social groups using a composite definition. If industrial workers experience only one or two of the three conditions of class specification – workers, for example, who have surplus value extracted from them and own no means of production but control their own labor process – are they less "authentic" members of the working class? What about individuals who are exploited in a capitalist labor process and also own a small business?[5]

Other problems of seemingly even greater moment arise when we

[5] These are similar to the problems that Wright's (1985) concept of "contradictory class locations" is able to address.

consider workers who are not involved in the production of surplus value at all – such as those involved in the distribution and exchange of commodities, or the circulation of finance and property, or those not in the paid workforce who labor in the domestic realm or voluntary sector. Many of these problems and questions associated with class analysis have arisen with greater urgency since recent episodes of industrial restructuring have appeared to complicate the social and economic terrain. As more and more people hold down two jobs, as more women enter the paid workforce, as work practices are changed to include the decision making input of workers, as the "informal sector" and hidden workforce increases, so traditional class mappings seem less and less relevant. Below we offer an alternative Marxian conception of class that may help to circumvent these conceptual and political dilemmas.

A different conception of class

As an alternative to layered and complex ways of defining class as a social *grouping*, we define class simply as the social *process* of producing and appropriating surplus labor (more commonly known as *exploitation*) and the associated process of surplus labor distribution (following Resnick and Wolff 1987). The importance of the relationship between our conception of class and the problematic to which we apply its analytics should not be underestimated. Our political and theoretical interest is in creating alternative (and potentially emancipatory) economic futures in which class diversity can flourish. Thus we are attracted to explicating class as a process and to highlighting its many different contemporary and potential forms. Others who undoubtedly share our emancipatory hopes are interested in class analysis for very different reasons such as, for example, a means of explaining (and contributing to eradicating) income and resource inequalities between social groups (Wright 1985: 65, 1993: 28). This problematic leads to a different conceptualization of class and its politics. In our view, neither approach is "right" or "wrong," but each class discourse has different implications and effects. We are interested in pursuing the class analytics developed by Resnick and Wolff because of the kinds of politics it helps us envision (which will become more visible by the end of this book).

By offering a "bare bones" definition of class as a social process of surplus labor appropriation and distribution, we hope to counteract the tendency to emphasize the social effectivity of property ownership, domination, and consciousness while ignoring exploitation. For a moment, then, we wish to hold exploitation up to the light and to analyze –

rather than presume – its relations to power, ownership, consciousness, and other social dimensions.

In *Capital*, Marx explored the specifically capitalist form of the class process, focusing in Volume I on the conditions of existence of capitalist exploitation (the production and appropriation of surplus labor in value form) and in Volume III on distributions of surplus value to various social uses and destinations. This dual focus enabled him to theorize some of the ways in which capitalist exploitation was both constituted by, and constitutive of, other aspects of social existence.

Marx's work generated a new and widespread understanding that processes of production frequently involve the performance and appropriation of surplus labor. Individuals produce more than is necessary to sustain them at a socially adequate level, and their "surplus labor" is appropriated, in a variety of forms, by other individuals and groups (sometimes including the performer of surplus labor her or himself).[6] We theorize contemporary processes of producing and appropriating surplus labor (class exploitation) as an aspect of social experience which is dimly and often unconsciously experienced and whose effects are often unrecognized and uncontested.

By producing a knowledge of exploitation as a social process, we hope to contribute to a more self-conscious and self-transformative class subjectivity and to a different politics of class activism and social innovation. Such a politics might not be concerned to eradicate all or even specifically capitalist forms of exploitation but might instead be focused on transforming the extent, type, and conditions of exploitation in particular settings, or on changing its emotional components or its social effects. It might not necessarily invoke the emancipatory agency

[6] The concept of surplus labor is highly abstract in the context of this general definition. Part of the theoretical work involved in any class analysis of a particular social site involves specifying the boundaries and other distinctions between necessary and surplus labor. Cameron (1995) and Hotch (1994) provide interesting examples of this type of theoretical endeavor. In Hotch's study of self-employment, which she theorizes as involving an "ancient" class process in which surplus labor is "self-appropriated" in the context of independent commodity production, the division between necessary and surplus labor is constituted as a moveable boundary, which shifts as the individual worker attempts to negotiate the difficult terrain of self-employment. If things are going well, the necessary component of the worker's labor may expand, as the worker enjoys a higher standard of living; but when things go badly, those means of subsistence may be cut back as the worker attempts to secure her continued existence as a self-employed worker by, for example, investing more money in her business. Thus, expenditures on child-care and recreation (representing the *necessary* or subsistence portion of the worker's labor) may decrease while expenditures on advertising and work-related phone calls (representing the *surplus* labor component, which is above and beyond the costs of reproducing the worker) may show a corresponding increase.

of a mass collective subject unified around a set of shared "interests" but could arise out of momentary and partial identifications between subjects constituted at the intersection of very different class and nonclass processes and positions (see chapter 8 and Gibson-Graham 1995a).[7]

Though concepts of class and exploitation have tended to be associated with capitalism, Marx and Marxists have identified a variety of noncapitalist forms of exploitation including ancient, primitive communist, feudal, slave, and communal class processes.[8] We distinguish, in addition, two distinctive moments within any class process, the *exploitative class process*, where surplus labor is produced and appropriated, and the *distributive class process*, where appropriated surplus labor is distributed to a variety of social destinations.[9] This proliferation and expansion of class categories facilitates the analysis of different forms and different moments of the class process, making possible the development of a complex knowledge of class and suggesting a range of noncapitalist class alternatives.

For us, creating a knowledge of class implies not only a concern

[7] This vision suggests the possibility of a complex rethinking of the social constitution and political potential of existing organizations that are currently identified with class politics (such as trade unions, business associations, or welfare rights groups) (see chapter 9 and Annunziato 1990, DeMartino 1991).

[8] Keeping in mind that each of these class processes is constituted in specific discursive and social contexts, so that it is impossible to provide a generic definition that will always "hold true," it might perhaps still be useful to offer an abstract description here of the class processes on this nondefinitive and openended list. Class processes are often distinguished from one another by the manner in which surplus labor is appropriated (Resnick and Wolff 1987: 309) and/or the way that one of the overdetermined circumstances of the appropriation has historically become associated with each. For example, in the ancient (or independent) class process a "self-employed" worker may appropriate her own surplus labor in value or use value form (see, for example, Hotch 1994). In a primitive communist class process the producers collectively appropriate their own surplus labor (Amariglio 1984). In a feudal class process the surplus labor of one individual or group is appropriated under conditions of fealty and mutual obligation in use value form, in return for the provision of means of subsistence. In a slave class process the surplus labor of the slave is appropriated by the master under conditions of servitude and the absence of freedom of contract. A communal class process involves the collective appropriation of surplus labor that has been produced by (some or all) members of the community. Finally, when we refer to capitalism as we prefer to understand it (rather than as it usually appears) we are referring to a class process in which surplus labor is appropriated from wage laborers in value form.

[9] This distinction is based upon that specified by Resnick and Wolff, who distinguish the "fundamental class process" of producing and appropriating surplus labor from the "subsumed class process" of surplus labor distribution (1987: 117–24). The different terminology we employ reflects our relative levels of comfort with the connotations of the two sets of terms, rather than an attempt to dissociate ourselves from the theoretical perspective of Resnick and Wolff.

about exploitation and economic difference but a commitment to an anti-essentialist theoretical position (see chapter 2). We do not wish to contribute to another Marxist knowledge that justifies itself by claiming that class is more fundamental or influential than other aspects of society and that, therefore, a knowledge of class has more explanatory power than other knowledges. Historically, such attempts to marginalize or demote other social processes and perspectives have created irresolvable conflicts and antagonisms between Marxism and other discourses of social transformation. At the same time, however, we do not wish to subordinate or subsume class to other social aspects such as power or property or consciousness or agency or struggle. We therefore theorize class as a process without an essence; in other words, class processes have no core or condition of existence that governs their development more closely than any other and to which they can be ultimately reduced.

We understand class processes as overdetermined, or constituted, by every other aspect of social life. By this we mean that we "think" the existence of class and of particular class processes by initially presuming overdetermination rather than by positing a necessary or privileged association between exploitation and some set of social processes (such as control over the labor process or consciousness or struggle or ownership, to rename the familiar few). In this initial presumption, class is constituted at the intersection of all social dimensions or processes – economic, political, cultural, natural – and class processes themselves participate in constituting these other dimensions of social existence.[10] This mutual constitution of social processes generates an unending sequence of surprises and contradictions. As the term "process" is intended to suggest, class and other aspects of society are seen as existing in change and as continually undergoing novel and contradictory transformations.

Theorizing class as an overdetermined social process rather than as a social grouping has certain implications for the nature and purpose of class analysis. Rather than involving the categorization of individuals and the disaggregation of societies into social groups, an overdeterminist class analysis examines some of the ways in which class processes participate in constituting and, in turn, are constituted by other social and natural processes. Class analysis theorizes society and subjectivity from the "entry point" of class, an entry point being an analytical starting place

[10] Such a presumption represents an epistemological choice on our part rather than an ontological commitment. We make no claims that class, or overdetermination, is implicated in the nature of being. But we are interested in operating within a discursive field in which essentialisms are not presumed as given in any sense. This reflects our political interest in creating space for thinking and enacting change in all social dimensions.

that reflects the concerns and preoccupations of a particular knower (Resnick and Wolff 1987). It yields a distinctive and partial kind of knowledge of the constituents and effects of class processes but does not accord explanatory privilege to the process of class.

Yet the process of producing an overdeterminist knowledge is itself contradictory. In a sense, the actual analysis of a particular class process involves the violation of the initial presumption of overdetermination. Examining, for example, the role of property law or of heterosexual norms and practices as conditions of existence of capitalist exploitation, we may come into conflict with the presumption of ceaseless change and transformation of each of these social aspects. We may posit for the moment processes that exist outside overdetermination (that is, not in change and not in contradiction) so that we can consider the ways in which these processes interact:

> Discourse is an attempt to freeze, to handle the ceaseless revolutions implicit in the concept of overdetermination, to do so by denying them in the fashioning of meaning . . . Discourse is an attempt to proceed *as if* – as if the objects it treats were secured, self-identical, reliable. (DeMartino 1992: 339–40)

Overdeterminist discourses cannot "reflect" overdetermination any more than essentialist discourses can correspond to the true state or essential nature of the world. But a form of social explanation that starts from the initial presumption of overdetermination will differ from one that starts from the initial presumption of essence (that is, from the founding presumption that a complex reality can be analyzed to reveal a simpler reality, an essential attribute, or a set of fundamental causes at its core). An overdeterminist discourse produces necessity (in the form of a determinate relationship between events or objects) as an effect of analysis rather than as an initial predication (DeMartino 1992).[11] In this way, causation/determination becomes a specific discursive effect rather than a pre-analytical ascription of ontological privilege.

[11] This distinguishes overdeterminism from, for example, critical realism, which incorporates into its epistemology the understanding that deep ontological structures participate in "generating social phenomena" (Bhaskar 1989: 3). In the terms of critical realism these underlying (or necessary) mechanisms that generate events and appearances are the "real" (Magill 1994: 115). In his considered discussion of the critical realists' preoccupation with deep structures and generative mechanisms, Magill offers a useful critique (though not from an overdeterminist perspective) of the resort in social sciences to a founding ontology. See also Barnes (1996: 15–23).

Alternative conceptions of social and individual identity and their implications for class politics

Like class defined as a social grouping, class defined as a social process is associated with particular ways of theorizing both society and political subjectivity. Through their distinctive treatments of these theoretical objects, the two ways of defining class yield very different implications for the nature and viability of class politics.

The Marxian conception of a class as the "conscious coming together of those who are similarly situated by production relations" (McIntyre 1991: 153) has historically been associated with images of industrial society as a centered unity. Society is typically theorized as a homogeneously or hegemonically capitalist formation centered on an industrial economy, with class theorized as a social relation originating in that center.

Perhaps because of its association with structural and systemic images of the social totality, "capitalism" in these conceptions tends to take up the available social space, incorporating the noneconomic dimensions of social life such as culture and politics as well as noncapitalist economic realms such as household production. Whether they are integrated with the economy through structural articulations, systemic logics, or hegemonic practices, these other aspects of society are colonized to some extent by the capitalist sector. Thus, social formations incorporating capitalist class processes are often theorized as *capitalist* formations, domestic labor is seen as *capitalist* reproduction, and the state and other institutions as implicated in *capitalist* regulation.

Such unitary and centered conceptions of capitalist society have fostered a conception of class as a (binary or expanded) structure founded in the relations of *capitalist* production. They have also given class struggle a leading role in social change. In a social "system" or coherent formation centered on an industrial capitalist economy, projects of class transformation are privileged sites of social transformation. The "working class" becomes the "subject" of history, the collective agent of fundamental change.

Because transformative efforts are seen as directed at systemic or hegemonic objects (for example, capitalist societies in their entirety), class transformation is often portrayed as a difficult, indeed, nearly impossible task. The politics of class transformation is enabled only at particular historical moments – usually those in which structural crisis (weakness) and working-class mobilization (strength) coincide. Given the heroic role it is asked to play, class struggle is often viewed as a military confrontation in which an army of workers is strategically deployed (Metcalfe 1991). Such a concerted and coordinated effort is required to confront the hegemonic unity of a coherent capitalist formation.

Social theorists have challenged images of social singularity (Laclau and Mouffe 1985) and, more particularly, the notion of a homogeneously capitalist society centered on an industrial economy, with its privileged role for working-class actors, its military metaphors of struggle, and its holistic conceptions of social transformation (of capitalist society into something else). Yet this vision, or elements of it, retains a degree of influence in political economy and other domains of social and cultural thought (see chapters 2 and 7). We are concerned to explicitly divorce class from structural or hegemonic conceptions of capitalist society because of the ways in which such conceptions discourage a politics of local and continual class transformation and make it difficult to imagine or enact social diversity in the dimension of class.

Concepts like the "mode of production" and the French regulationists' "model of development" tend, by virtue of their structural integrity, to confer unity and stability on otherwise amorphous social formations (see chapter 7). At the same time, by virtue of the centrality they accord to production, they identify the politics of class as the "structural politics" of the singular social. In the familiar script for class politics, the unified and coherent society can be ameliorated and reformed through everyday political activities but can only be transformed through systemic upheaval (for example, the breakdown of a social logic or structure in the context of a coordinated transformative struggle).

In order to liberate class politics from these restrictive yet privileged scenarios, we wish to understand society as a complex disunity in which class may take multiple and diverse forms. Primitive communist, independent, slave, feudal, capitalist, and communal class processes can, and often do, coexist. In this conception, then, an "advanced" industrial social formation is not a coherent and stable unity centered on capitalist class relations. It is a decentered, fragmented, and complexly structured totality in which class and other processes are unevenly developed and diverse.

An industrialized social formation may be the site of a rich proliferation of class processes and a wide variety of class positions – producer, appropriator, distributor, or receiver of surplus labor in a variety of forms. Class processes are not restricted to the industrial or even the capitalist economy. They occur wherever surplus labor is produced, appropriated, or distributed. The household is thus a major site of class processes, sometimes incorporating a "feudal" domestic class process in which one partner produces surplus labor in the form of use values to be appropriated by the other (Fraad et al. 1994).[12] The state may also be a site of exploitation, as may educational institutions, self-employment, labor unions, and other sites of production that are not generally associated with class.

Because class is understood as a process that exists in change, the class "structure" constituted by the totality of these positions and sites is continually changing. Projects of class transformation are therefore always possible and do not necessarily involve social upheaval and hegemonic transition. Class struggles do not necessarily take place between groups of people whose identities are constituted by the objective reality and subjective consciousness of a particular location in a social structure. Rather, they take place whenever there is an attempt to change the way in which surplus labor is produced, appropriated, or distributed.

Classing Sue and Bill

When the systemic representation of a homogeneous capitalist social formation is replaced with the alternative conception of a decentered and complex heterogeneity, class like other processes becomes visible as heterogeneous and unevenly developed. Independent, communal, capitalist, slave, feudal, and other class processes – obscured by the conception of a singular and systemic social identity – can be acknowledged and theorized as constituents of contemporary social formations.

Like the "identity" of social formations, individual class identity can be understood as decentered and diverse. Individuals may participate in various class processes, holding multiple class positions at one moment and over time. To exemplify the notion of multiple and fractured class identity at both the personal and social levels, we recount below the stories of a Philippines-born nurse, Sue, and her white Australian coal miner husband, Bill.

Sue and Bill met while Bill was recuperating from a football injury in a large Brisbane hospital several years after Sue migrated to Australia as a trained nurse. They married and sometime later moved to rural Queensland to take advantage of jobs opening up (for men) in the modern "open cut" mines. Bill is currently a coal hauler, a job for which he has no formal qualifications. He is defined by the company as a wages worker (in other words, he plays no supervisory role). With overtime and weekend

[12] The use of the term "feudal" in the household context has provoked controversies that are explored in chapter 9.

work, he earns about $65,000 per annum, well above what truck drivers in other industries earn and on a par with senior university professors. At his mine, a profit-sharing scheme has been introduced to encourage productivity gains and discourage industrial disruption, so Bill receives an additional payment to complement his wage. Bill has saved part of his income and invested his savings. He owns a block of rental units on the coast and a portfolio of shares in leading companies operating in Australia (some productive, others financial).

As a country boy by origin, Bill is a strong supporter of the conservative National Party, but as a coal miner he is required to belong to the United Mineworkers Federation of Australia. This industrial union is one of the more militant worker organizations in the Australian economy. In his time off from work, Bill runs his own small business shooting wild pigs and arranging for them to be frozen and shipped to market. His role at home is very traditional. The only domestic work Bill regularly performs is keeping the yard tidy, putting out the garbage, and driving the kids to sporting activities on the weekends. Bill keeps tabs on his income and allocates a weekly portion to Sue for housekeeping purposes and a monthly amount to be sent to her family who own a small business back in the Philippines.

If we understand class as a social group, Bill's class location is difficult to ascertain. He is a wage laborer from whom surplus value is derived. He has little control over his own labor process. Yet he owns shares in productive capitalist enterprises and receives a small share in the profits made by the mining company that employs him. He is a member of two political organizations with quite antithetical philosophies and is active in both. If we were to give priority to his role in the relations of production at the mine, we might be tempted to see his actions in the Mineworkers Federation as being in his "true" class interests, explaining his participation in the National Party as a product of a "false consciousness." But if we were to give priority to his relations of production as a pig shooter or to his distribution of funds to another small business overseas, the opposite might apply. Bill's membership in the "working class" can only be secured by emphasizing some of the relations in which he participates and de-emphasizing others: that is, by ranking the components of his experience in a hierarchy of importance, or by reducing his total social experience to a set of fundamental or essential elements.

Sue's story is even more difficult to tell in traditional class terms. Sue is a Filipina and a trained nursing sister who had to give up paid employment when she moved with Bill to the remote mining town. She is now a full-time carer for her husband and three children. She shops for provisions, produces food, clean clothes, and an orderly and comfortable environment, and is the primary manager of family relationships with friends and service providers. In addition she takes some financial responsibility for the education and welfare of her extended family members in the Philippines and is active as a volunteer social worker in the local support group for Filipina wives of Australian miners. In fact, Sue is the classic multiskilled flexible worker. Her hours of work are usually longer than Bill's. When Bill is on day shift, she rises to cook his breakfast at 5 a.m., and when he is on afternoon shift, she irons and does other chores in the evening while waiting for him to come home at midnight. When Bill goes off pig shooting on weekends, Sue takes over his parenting role, driving children to sporting events and supporting their leisure activities. Bill and Sue have a joint bank account and they jointly own the block of rental units on the coast. At the same time Sue is dependent on her relationship to Bill for access to the company-owned house in which they live and for the means of her domestic production. She is not a member of a political party but votes for the National Party along with her husband. While she is from a more economically privileged background and a different language group than many other Filipina women in the town, Sue identifies with this group, under the pressure of the inadvertent as well as outright racism of her "host" society.

If we emphasize Bill's role in the construction of Sue's class identity, she might be seen as a member of the working class by virtue of her marriage, her reproductive role, and her allegiance to the wage-earner social set. Before she moved to the mining town, however, she was employed as a nurse in a supervisory position that distinguished her from those with no control over their own labor or that of others. And before she migrated to Australia she was the relatively well off daughter of a member of the petit bourgeoisie in the Philippines. Given that one of these "class locations" belongs to her husband, one to her past, and one to her father, class as a social category would not seem directly relevant to an understanding of Sue.

From a class process perspective, however, it is not difficult to ascertain Sue's class position, nor is it necessary to ignore or demote any of Bill's experiences in order to "place" him with respect to class. Sue is engaged in a "feudal" exploitative class process of surplus labor production and appropriation in her role as wife and mother in the household (see chapter 9 and also Fraad et al. 1994). Her labor is appropriated by Bill in return for the provision of shelter and access to the commodified means of domestic production. A host of cultural, familial and companionship practices also provide conditions of existence of Sue's exploitation, as does the widely held view that Bill "brings in the money" and therefore deserves to be served and sustained. Sue is also involved in a volunteer class process in which her surplus labor is appropriated by members of the Filipina group in town. A host of discourses about race, dependency, solidarity and national loyalty provide conditions of existence of this form of exploitation. Sue's lack of public political involvement is influenced by her participation in both these class processes as well by a whole set of other processes, including the social construction of a racialized "Asian" identity and an "emphasized" femininity[13] and motherliness in the unique culture of the mining town.

From this perspective, then, Sue is engaged in two class processes that are involved in constituting her complex subjectivity and overdetermined by her role in the traditional gender division of labor and by the racialized construction of "otherness." When her experience is theorized in this way, her occasional struggles with Bill over his performance of household chores can be seen as struggles over the degree of exploitation she is willing to accept at home. In other words, Sue is not unconcerned with class, nor is she apolitical as her husband and his union mates think. Rather, she is intermittently involved in a non-solidary politics of class.

While a class process approach makes visible the role of class in constituting Sue's subjectivity, it allows us to theorize Bill's class identity without giving priority to one aspect of his activity and experience. Bill performs surplus labor that is appropriated from him in the capitalist exploitative class process as surplus value. He also receives a distributed profit share as well as dividend checks, thus participating in a capitalist distributive class process.[14]

As a self-employed pig shooter and marketer, he is involved in an independent class process in which he produces, appropriates and distributes his own surplus labor. In the home, he appropriates the unpaid labor of his wife in a feudal exploitative class process that is a familiar if not dominant constituent of the contemporary Australian household. The political processes in which Bill participates are influenced by his participation in all these different class processes as well as by other processes, like the social construction of hegemonic masculinity,[15] white supremacism and the ideology of

[13] Defined by Connell (1987: 183) as involving compliance with the subordination of women to men and accommodation to the interests and desires of men.

[14] Bill's positioning as a recipient of distributed surplus value helps to ensure the smooth functioning of coal extraction and softens the impacts of his alter positioning as an exploited worker.

solidarity among the working class. Like the social formation of which he is a part, Bill is a contradictory and fragmented social site, the intersection of many different class (and nonclass) processes. No one of these processes defines his true identity or his "class" interests, though each participates in doing so, along with many other class and nonclass processes.

To search for Bill's true and singular class identity in this complex and shifting intersection would involve a quest for the type of "regulatory fiction" that Butler (1990: 339) sees gender coherence to be. Both rely on a conception of identity as singular, homogeneous, and fixed rather than multiple, fragmented, and shifting.[16]

The involvement of Sue and Bill in a variety of different class processes has changed over time and overlapped in different ways. In their adult lives, however, they have never not been engaged in class processes. In fact, these involvements have participated in constituting them as acting and powerful subjects in many political arenas, both publicly (at the mine site, the hospital, the school, or the community center) and privately (in the home and family).

Because their relationships to class have not constituted them as members of a particular social group (the "working class," for example), a class analysis of Sue and Bill does not threaten to subsume or subordinate their identities as gendered or ethnic or otherwise differentiated subjects, nor does it necessitate positing a unified class identity. This by now familiar decentered approach to identity and class has a variety of implications for understanding class politics, some of which we explore below as we examine several dimensions of women's involvement in the politics of class.

Women, households, and class

Largely because class debates have concentrated upon the definition and nature of *capitalist* classes, women have often found themselves in a

[15] Connell (1987: 183) notes that "hegemonic masculinity" is always constructed in relation to subordinated masculinities as well as in relation to women. In mining communities, where there is very little tolerance for any subordinated masculinities, the very hegemony of a certain masculine identity is of course itself constituted by class and geography.

[16] In Butler's (1990: 337) words, "If the inner truth of gender is a fabrication and if a true gender is a fantasy instituted and inscribed on the surface of bodies, then it seems that genders can be neither true nor false but are only produced as the truth effects of a discourse of primary and stable identity."

problematic position *vis-à-vis* class. Those women not in the capitalist labor force have assumed class positions only through their relations to others (usually husbands or fathers).[17] The wives of workers, for example, are often considered vicarious members of the "working class." Their intermittent, muted, or absent class militancy has been attributed to their socialization as unassertive "conservators" or as denizens of a private sphere with "backward" political and social concerns.

As workers in the domestic economy, women have often been theorized as engaged in a nonclass process of "reproducing" the capitalist workforce – feeding, clothing, nurturing, cleaning – performing a socially necessary (if hugely undervalued) function:

> the home is a key site of the day-to-day and generational reproduction of labour-power (which) is oriented towards fulfilling the needs of capitalist production. (Mackenzie and Rose 1983: 159)

Many authors have characterized the home as a separate "sphere" of reproduction and consumption. Here domestic labor (largely performed by women) organizes the consumption of the commodities produced in a capitalist labor process. Women's work in service to capitalism is performed under the governance of patriarchy, a system of rules and practices of gender domination.

Certain feminists have attempted to redress the secondary status of the reproductive or patriarchal sphere, which seems ordained to serve capitalism just as many women serve men. Arguing against the tendency to subordinate patriarchal oppression to class exploitation, these theorists have generated a "dual systems" approach that takes a variety of forms. In one form of dual systems theory, patriarchy is transported into the sphere of class. Capitalism and patriarchy are viewed as two systems of social relations that interact in every domain of social life; gender relations are thus "*part of* the 'relations of production'" (Connell 1987: 45) and patriarchy cannot be relegated to a separate sphere.

In a second form of dual systems theory, class is transported into the sphere of patriarchy. The household is theorized as the locus of a patriarchal or domestic "mode of production" (see, for example, Folbre 1987; Delphy 1984; Delphy and Leonard 1992) that functions according to a logic distinct from that of the capitalist mode of production but

[17] In categorizations of social class such as that of the British Registrar General, married women and widows are classified according to the occupations of their husbands (Krieger and Fee 1993: 68). This assumption of homology has carried over into views on political subjectivity. Pratt and Hanson (1991: 245) argue that "traditionally, women's class consciousness has simply not been theorised; it was presumed that a woman's subjectivity (at least that related to class) could be 'read off' that of her relevant patriarch (husband or father)."

that interacts with the latter to constitute the social whole. Most of these conceptualizations have foundered on the difficulty of theorizing patriarchy as a system or social structure (Connell 1987: 46).[18] If patriarchy is a mode of production, what are its laws of motion and how do they interact with those of capitalism? Such questions have proved quite intractable. Theorists have also been concerned about what may appear as an attempt to colonize the household as a domain of class. Given the history of feminist struggles against the totalizing ambitions of traditional Marxism, feminist analysts have tended to theorize patriarchy and gender domination as socially pervasive while giving class a more restricted domain (Connell 1987).

It seems to us, however, that excluding class from the household has the effect of making invisible the production, appropriation, and distribution of surplus labor (and the struggles over these class processes) that go on in that particular social site. We would therefore like to renegotiate the relationship between class and the household, divorcing household and other class processes from the idea of economic and social "systems" or structures. Our intention is not to displace or replace gender relations as a category for household analysis. Instead, we wish to add the dimension of exploitation or class.

In our understanding, industrial social formations are the sites not just of capitalist class processes but of noncapitalist class processes as well. The household, then, can be seen as involved not only in capitalist reproduction but also in the reproduction of noncapitalist class processes such as the independent class process of self-employment and "self-exploitation" (Gabriel 1990; Hotch 1994). More importantly for our purposes, it can be constituted as an autonomous site of production

[18] Not to mention the conceptual problems associated with the potentially infinite replication of "systems" that structure different types of social advantage. Alluding to the "enormous . . . political impact of black women's critique of the racist and ethnocentric assumptions of white feminists," Barrett and Phillips note, for example, that "the social structural models of society that had been organized around the two systems of sex and class found a third axis of inequality hard to accommodate; the already acute difficulties in developing a 'dual systems' analysis were brought to a head with the belated recognition that ethnic difference and disadvantage had been left out" (1992: 4).

It is interesting to note that despite Connell's awareness of the theoretical difficulties attending the multiplication of social structures, in his recent and exciting book *Masculinities*, he constitutes an "ontoformative gender order" which displays tendencies toward crisis (1995: 83–4) and in which a variety of masculinities and femininities exist as "configurations of gender practice." The purpose of this (from our perspective) very worthwhile but problematic theoretical attempt to proliferate and differentiate masculinities is ultimately to provoke "a transformation of the whole (gender order) structure" (1995: 238).

in its own right in which various class processes are enacted. And according to Fraad et al. (1994) it has become an important zone of class conflict over the past 25 years.

The (white) heterosexual household in industrial social formations has often been a locus of what we have called a feudal domestic class process (see chapter 9), in which a woman produces surplus labor in the form of use values that considerably exceed what she would produce if she were living by herself. When her partner eats his meals, showers in a clean bathroom, and puts on ironed clothes, he is appropriating her labor in use value form. Throughout much of the twentieth century, this form of exploitation has seemed fair and appropriate because the man generally worked outside the household to procure the cash income that was viewed as the principal condition of existence of household maintenance.[19] Even when it did seem unfair (in cases, for example, where the woman worked outside the home or where it was recognized that some women worked longer hours and that they had no vacations and were not permitted to retire), the lack of alternatives for women often kept them from attempting to transform their class positions. Familiar cultural presumptions about the natural or divine origins of women's household role had a similar effect.

During the past 25 years, however, women's household exploitation has increasingly been seen as unfair and as something to be struggled over. This change has to do with many things, including the second wave of feminism and heightened feelings of equality and commensurability brought about by women's greatly increased participation in waged work (not to mention the increased pressures on their time). For a variety of reasons, then, some women now feel entitled to an equal domestic load and to a democratic decision-making process about the allocation of the various types of domestic work. In many households, the issue of household-based exploitation is on the table and the feudal domestic class process is in crisis (Hochschild 1989; Fraad et al. 1994).

Many feudal households have broken up and their members have re-established themselves in independent households as "self-exploiting"

[19] Increasingly research into the domestic situations of families of different races, classes, sexualities and localities over the course of industrial history is pointing to the atypical nature of the one breadwinner household. From our perspective this research highlights the multiple class processes that always have and continue to take place in households despite the dominance of a particular discourse around the normality of the male breadwinner and female domestic carer (Delphy and Leonard 1992: 132–3).

performers and appropriators of domestic surplus labor.[20] According to a study by Burden and Googins (1987), establishing an independent household is a way for some women to achieve a reduced work week. Married mothers in this study spent 85 hours per week on job and family responsibilities, while single mothers spent 75 hours per week on these tasks. Other households have instituted communal class processes whereby all members perform surplus labor and jointly appropriate it, democratically allocating both work and the fruits of work (Kimball 1983). In heterosexual communal households, which are less likely than feudal ones to be structured around the priority of the male's career, difficulties may arise over how to challenge traditional gender roles that undermine communalism. Men, for example, may confront the loss of public and private status associated with being the higher wage earner as they opt for more flexible working hours that allow an equal role in child-care and domestic labor. Women confront the mixed emotions associated with relinquishing the role of primary caregiver to children and quality controller over household cleanliness and atmosphere.

The communal household is not without class antagonism and conflict. Indeed, the negotiation of work and space, both physical and personal, may be more difficult under its rather experimental conditions.[21] Clearly the historical difficulties of the feudal household and more recent problems with establishing communal households have contributed to the accelerated growth of independent households where class and gender conflicts are resolved through the establishment of solo householding. In these households, there is no gender division of labor to negotiate and the adult householder is in sole charge of the production, appropriation, and distribution of her/his own surplus labor.

"Restructured" households are often seen as the outcome (or casualties) of struggles against patriarchy and gender oppression. We might see them as also the outcome of struggles around class. Though they have not

20 See Cameron (1995) for detailed case studies of independent forms of surplus labor appropriation and distribution in single and multiple-resident households.

21 In a fascinating study of gender identity and class as a process Cameron (1995) describes the case of a marriage that broke up over the gender implications of a communal class process. In this household the husband began to assume a greater responsibility for domestic labor upon retirement. His wife relinquished her command of the kitchen, shopping and other tasks and continued to work in the paid workforce. She found that her loss of a domestic subjectivity (especially the more public role as cook) was too much to bear. The loss of femininity and power that it signified forced her out of the household, ultimately out of the marriage and into a solo domestic situation where her gender identity is being reshaped around an independent domestic class process.

articulated their goals in the language of class, many women have become uncomfortably conscious of their exploitation in the household as well as of their gender oppression in the same domain (Hochschild 1989). Gendered class struggles over the performance, appropriation, and distribution of surplus labor have contributed to the growth of households where communal class processes are in place (Kimball 1983) and to the more rapid rise of independent households where class conflicts have been resolved through gender separation.

This narrative of transition in household class relations in industrial societies has a number of distinctive elements. Rather than being seen as governed by a hegemonic structure or set of rules like a patriarchal mode of production, the household is represented as a social site in which a wide variety of class, gender, racial, sexual and other practices intersect. Because this site is not subsumed to an overarching and stable social system (capitalism or patriarchy being the usual suspects) it can be theorized as a locus of difference and constant change. Each local instance is constituted complexly and specifically, unconstrained by a generic narrative or pattern from which it may only problematically stray.

In the context of systemic or hegemonic social representations, a local politics of "resistance" is often portrayed as relatively powerless in relation to the hegemonic structure; even when struggles are deemed successful, their successes may be negated on the theoretical level where the "system" or hegemonic formation is reasserted fundamentally untouched. By contrast, in the world of our narrative, where class is not constituted within a social structure, class politics in the form of individual struggles over exploitation is an ever-present experience with significant (though not unidirectional) transformative effects.

This is not to say that the narrative of crisis in the feudal household, which provides a single story of household development and differentiation, is not itself a form of hegemonization. It could certainly be displaced by other representations in which households were always already differentiated from each other (making the feudal household visible, say, as a white middle-class fantasy or regulatory fiction). Cameron (1995) has recently told a number of household stories in which each household comprises many different class processes at any one moment, and these interact with gender, sexual and power relations in contradictory ways (rather than lining up quite so neatly as they do in the narrative above). In Cameron's representation, household exploitation and heterosexual gender difference lose the negative taints that have accrued to them through their longstanding associations with hegemony, domination and oppression, becoming visible as highly differentiated, quite fragmentary and continually under renegotiation.

But, of course, how one chooses to represent a social site has to do with any number of things, including the politics that one is interested in promoting.[22] The story of "*the* crisis" in "*the* feudal household" may speak to political subjects already constituted by existing feminist discourses as concerned with inequalities in household labor, while Cameron's work is more oriented to promoting an alternative vision of female subjectivity and agency as a counter to prevalent (feminist) theories that constitute women as dominated, devalued or oppressed and as only powerful if they are engaged in specific practices of resistance, or resistance to specific practices.

Conclusion

The notion of the "working class" as the collective subject of history can be seen as the effect of Marxist and non-Marxist discourses about the principal and defining role of industrial capitalism in structuring developed western social formations. These discourses of "capitalist development" have fostered a conception of society as structured by two major classes defined objectively by capitalist relations of production and subjectively by the political and cultural experience of industrialization.

The discourse of class which has depicted class as the central social relation of contemporary societies is now contributing to its marginalization. Critics of Marxism proclaim the death of class, while Marxist theorists of contemporary capitalism lament working-class demobilization. From our perspective, what has died or been demobilized is the fiction of the working class and its mission that was produced as part of a hegemonic conception of industrial capitalist development. As this conception has been devalued by criticism and other historical processes, and as multiple social "centers" and contending forces have seized the historical stage, the "working class" has been peripheralized and demoted. Discursive moves to displace the economic essence of society have displaced as well its agents of transformation. Now the militaristic image of a massive collectivity of workers all defined by a similar relation to industrial capital is part of a receding social conception and politics of change.

Despite the waning theoretical and political fortunes of the "working class," class itself may still be theorized as present and pervasive. Monolithic images of the "working class" associated with craft unionism and Fordist industries may no longer be recognized by social theorists or those who labor. They may not work to mobilize resistance and impulses

[22] See, for example, chapter 9 for a conceptualization of feudal households that is overdetermined by union and community politics in Australian coal-mining towns.

toward social transformation or play a leading role on the stage of social theory. But class is not thereby necessarily diminished as an intelligible constituent of social development and political change. Instead, the role of class as a social process may be recast in different social and theoretical settings, ones in which new political opportunities may emerge.

For us, the question today is not whether class is a concept with continuing relevance, for discourses of socioeconomic differentiation[23] and surplus labor appropriation are still, and perhaps will always be, involved in the constitution of social knowledges and political subjects. Our question is, how can theorizing class as a process of production, appropriation, and distribution of surplus labor add dimensions to theories of society and to projects of social and economic innovation?[24] How may it contribute to conceptualizing and constituting decentered and multiple selves that are always in some ways political (powerful) subjects (Kondo 1990)?

A view of social subjects as multiply constituted by class processes as well as other social processes does not allow us to presume certain "class interests" or "class capacities," nor does it lead to a theorization of likely "class alliances." At the same time, it does not preclude the envisioning of collective action. In the alternative space we see for a politics of class we may encounter and even foster the partial identification of social subjects around class issues and the formulation of strategic solidarities and alliances to effect class transformation. Importantly though, we are always aware that these solidarities are discursively as well as nondiscursively constructed and that a class "identity" is overdetermined in the individual social subject by many other discourses of identity and social differentiation.[25] This conception of class also allows us to see many non-class-oriented social movements as having profound effects on class transformations, possibly liberating the potential for the

[23] In the emerging discourse of social polarization, for example, class has become a prominent social descriptor, as images of a growing "underclass" and shrinking middle class proliferate in various discursive settings.

[24] In a forthcoming edited collection (Gibson-Graham et al. 1997) we present a number of different class analyses that may contribute to imagining alternative class futures.

[25] Many of the "failures" of class politics have grown out of an inability to recognize subjects as positioned in gendered or racialized discourses as well as multiple class discourses, and the tendency to ignore the sometimes contradictory overdeterminations between these discourses. We are thinking here of the very problematic relationship that has existed between the formal labor movement and women and minorities. Historically, in solidary movements of all kinds, there has been a tendency to theorize sameness as the basis of unity and solidarity, with a consequent denial or elision of difference that has had problematic and divisive effects. As Hotch notes, "theorizing unity instead of difference has an effect, but that effect may not be unifying" (1994: 26).

development of class diversity in ways that targeted "class politics" has not.[26]

Our purpose here is not to create a "better" form of knowledge or one that will lead to a "better" politics of change. We are interested in producing a class knowledge that is one among many forms of knowledge and not a privileged instrument of social reconstruction. But we also have an interest in posing alternative economic futures. Towards this end we argue that a new knowledge of class may contribute to a revitalized politics of class transformation.

[26] For example, the women's movement and the environmental movement have generated new discourses of social and ecological identity that have had major impacts upon exploitative class processes in the household and distributive class processes within the enterprise (see chapters 8 and 9). There are ways, then, that we can identify these movements as also class movements without claiming them as such. Indeed, thinking about the impacts of the various strategies of these movements from a class entry point is a useful way to start building points of collective identification and group action. Sullivan (1995) urges feminists, for example, to specify the differences between forms of prostitution practiced under varying legal, economic, racialized and discursive relations rather than simply adopting universalizing positions about the "wrongness" of prostitution – with all the familiar accompanying assumptions about women who work as prostitutes. An analysis of the many different class processes operating within the sex industry (see, for example, that of van der Veen 1995) and their conditions of existence would be an interesting place to begin building collective feminist and class-oriented actions.

4

How Do We Get Out of This Capitalist Place?

> Writing has nothing to do with signifying. It has to do with surveying, mapping, even realms that are yet to come.
>
> (Deleuze and Guattari 1987: 4–5)

Geographers – of which I am one (or, more accurately, two) – have ambivalent feelings about the proliferating references to space in contemporary social theory. And, indeed, the profusion of spatial metaphors is remarkable (as well as frequently remarked). Discursive space is "occupied," speaking positions are "located" or "situated," "boundaries" are "transgressed," identity is "deterritorialized" and "nomadic." Theory flows in and around a conceptual "landscape" that must be "mapped," producing "cartographies" of desire and "spaces" of enunciation. If space is currently where it's at, perhaps it is not surprising that professional geographers occasionally feel displaced. It seems we are all geographers now.

The spatialization of theoretical discourse owes something to structuralist theories in which linguistic or social elements are seen as defined relationally, via a "synchronic" articulation. But it is usually attributed more directly to poststructuralists like Foucault (see, for example, 1980) and Deleuze and Guattari (1987) as well as to "pre-poststructuralists" like Althusser and Gramsci, and the Marxian tradition to which they belong.[1] Indeed, the spatial metaphors associated with Marxian analysis – "colonization," "penetration," "core and periphery," "terrains of struggle" – are not dissimilar to those of poststructuralism.[2] Both types of theory represent space constituted by or in relation to "Identity"[3] or Form. While poststructuralist theory is concerned with problematizing the fixing of Identity and tracing the performance space of multiple and fluid identities, Marxian theory has generally been focused upon the performance space of one type of Form – the mode of production or, more particularly, Capitalism.

After struggling for so long to erect and strengthen the ramparts of an academic identity in the shadow of more established disciplines, geographers now find all sorts of strange beings camped outside or scaling the battlements eager to assume the language of geography, if not to take up positions in its defense. For one who has dwelt protected within the disciplined space of geography, this invasion is welcome. Indeed, it is the wordy invaders who have kindled in me, for the first time, an interest in "space" – a core, even foundational, concept within my professional dwelling place. But while "we" all might be geographers – or at least explorers – now, some disciplinary geographers (despite feeling partially vindicated) are worried.

Massey (1993: 66) is concerned that the proliferation of metaphorical uses of spatial terms has blurred important distinctions between different meanings of space. And Smith and Katz are alarmed at the use of spatial metaphors in contemporary social and cultural (not exclusively poststructuralist) theory that take as their unexamined grounding a seemingly unproblematic, commonsense notion of space as container or field, a simple emptiness in which subjects and objects are "situated" or "located." These metaphorical attempts to contextualize, relativize and de-universalize social sites and speaking positions inadvertently invoke a standpoint at a set of coordinates, a location in a naturalized and asocial "absolute" space. Yet the very conception of absolute space, they caution, is *itself* socially produced and historically specific: the representation of space as an infinite, prior and neutral container or grid, in which discrete entities operate independently of one another and of space itself, gained ascendancy with the philosophers Newton, Descartes and Kant and was "thoroughly naturalized" with the rise

1 See Smith (1984) for a discussion of Marxian references to space.

2 Ferrier (1990) points out that spatialization and, in particular, cartography are actually central to modernist forms of representation and subjectivity, and instrumental to modernist projects of subjugation. She argues that precisely for this reason cartography has become an important metaphor within contemporary projects of rethinking.

3 Here I refer to Identity (or the Idea) in the symbolic domain, implying the quest for ultimate definition or for the fixing of signifier to signified. In this chapter, as in this book, I move between three "types" of identity in order to develop their overlaps, connections, and contradictions: *Identity* as defined above, the *identity* of "the social" as a complex totality (often referred to as society and sometimes divided into culture, economy and polity) and individual *identity*, that which constitutes subjectivity and agency. In this chapter I play with these three senses of identity, slipping between and among them with barely a warning. I hope that this little game will be more productive than confusing for the reader.

of capitalism between the seventeenth and nineteenth centuries (Smith and Katz 1993: 75–6). By proliferating spatial metaphors without problematizing the representation of space, social theorists reproduce a view of space that is "politically charged in its contemporary implications as much as in its historical origins" (p. 76).[4] An understanding of space as a coordinate system in which locations are clearly defined and mutually exclusive has contributed, for example, to an "identity politics (that) too often *becomes* mosaic politics" (p. 77, emphasis theirs), that is, a politics of competition and fragmentation.[5]

In *The Production of Space* Henri Lefebvre, the Marxist theorist of space and spatiality, expresses a related but more extreme disapproval of the appropriation of spatial metaphors by philosophers, especially poststructuralist ones:[6]

> Consider questions about space, for example: taken out of the context of practice, projected onto the place of a knowledge that considers itself to be "pure" and imagines itself to be "productive" (as indeed it is – but only of verbiage), such questions assume a philosophizing and degenerate character. What they degenerate into are mere general considerations on intellectual space – on "writing" as the intellectual space of a people, as the mental space of a period, and so on. (Lefebvre 1991: 415)

Suspecting the dissociation of conceptual space from "lived" space (which he identifies as a pre-discursive terrain of production), Lefebvre sees philosophers' production of mental space as only "apparently extra-ideological" (1991: 6). While poststructuralist theorists might imagine

4 Smith and Katz see it, paraphrasing Lefebvre, as a "conception of space appropriate for a project of social domination," one that "expresses a very specific tyranny of power" (p. 76). In Rose's (1996) reading, Smith and Katz argue that "spatial metaphors which refer to absolute space are regressive because absolute space serves to freeze and thus to sanction the socio-spatial or theoretical status quo" (pp. 2–3).

5 This depiction bears some affinity to Laclau and Mouffe's (1985) discussion of the essentialism of the fragments. When the structural essence of the social (e.g., the capitalist mode of production) is discursively displaced by a heterogeneous multiplicity of social sites and practices – as, according to Laclau and Mouffe, it is in the work of Hindess and Hirst – the essentialism of the totality is effectively replaced by an essentialism of the elements; in other words, each part of the "disaggregation" takes on a fixed and independent identity rather than being relationally defined. In an overdeterminist Althusserian conception, by contrast, "far from there being an essentialist *totalization*, or a no less essentialist *separation* among the objects, the presence of some objects in others prevents any of their identities from being fixed" (p. 104).

6 In particular, he was concerned with the spatial language of Foucault, Derrida, Kristeva and Lacan and their promotion of "the basic sophistry whereby the philosophico-epistemological notion of space is fetishized and the mental realm comes to envelop the social and physical ones" (1991: 5).

themselves to be undertaking transgressive acts via their work, Lefebvre remains convinced that this work, by detaching mental/conceptual space from social/material space, unwittingly reproduces the "dominant ideas which are perforce the ideas of the dominant class" (p. 6). The familiar implication here is that only through the dialectic of practice and reflection, that is, at the intersection of language and social action, will true (read revolutionary) spatial and social understandings be produced.

While Lefebvre berates philosophers and cultural theorists for their failure to recognize the lived materiality of space as the appropriate basis of all discursive representations in mental space, Smith and Katz warn against the failure to situate spatial metaphors in an historical materialist (and therefore relative) frame. Together they are concerned that discursivity and materiality be made to touch lest spatial metaphors be rendered complicit in capitalist reproduction.

For these Marxists and geographers alike there appears to be a concern that the materiality, sociality, and produced nature of space might be ignored by those who so readily employ spatial metaphors in poststructuralist discourse. Their concern is traced to the worrying political implications of somehow disregarding "reality."[7] Without a true grounding in the material social world, they wonder, how can spatial representations become appropriately (rather than regressively) political?[8]

As battles between metaphor and materiality, discourse and reality rage in and around us, and "the enemy" infiltrates our disciplinary boundaries, what better time might there be for a jump into space? An engagement with space allows us to confront some of the political and epistemological concerns about the relationship between discursivity, materiality, and politics that have arisen in the clashes between modern and postmodern feminist and urban discourses. It opens up possibilities of thinking from the outside in, both from the poststructuralist encampments into the protected dwelling of geography, and from the space of formlessness into the space of Form: "The outside insinuates

[7] In Rose's view, there is for all these geographers "a real space to which it is appropriate for metaphors to refer, and a non-real space to which it is not" (1996: 3).

[8] This is a question that is interestingly parallel in structure to one often posed to feminist poststructuralists: without a true grounding in the materiality of women's experience, how can poststructuralist feminist theoretical interventions avoid functioning in service of a dominant masculinism? In the current context it is the dangers of fragmentation (for the left and for feminism) that are seen to open "us" up to the enemy – revealing the modernist vision of solidary resistance and organization that provides the foundations of this critique. As a geographer and feminist not overly worried by the prospect of fragmentation (see Gibson-Graham 1995), I am of course not alone. Soja and Hooper, for example, welcome the proliferation of discursive spatialities and the new "postmodernized and spatialized" politics of difference (1993: 184).

itself into thought, drawing knowledge outside of itself, outside what is expected, producing a hollow which it can then inhabit – an outside within or as the inside" (Grosz 1994: 9).

By examining the spatial images that have been employed in feminist analyses of the body and the city, we may trace the political effects of privileging the materiality of women's experience and capitalist social relations. At the same time, we may discover some of the political potential of an alternative conceptualization in which discourse and other materialities are effectively intertwined.

Rape space, modern space

Recent feminist theorizations of the body employ and also challenge the familiar spatial language of "inside/outside," "surface/depth," "emptiness/fullness," "dwelling." Spatial knowledges of women's bodies and female sexuality have of course both philosophical and activist origins. For the moment, I would like to explore feminist knowledge of the body gained through women's activism around rape.

The prevailing (though not exclusive) feminist language of rape situates it as a fixed reality of women's lives – a reality founded upon the assumed ability of the (male) rapist to overcome his target physically (Marcus 1992: 387). Creating a public knowledge of rape as a "reality" has been one of the projects of anti-rape activists and policymakers, and making rape visible in the community constitutes a significant victory for feminist politics.

Sharon Marcus is a feminist who challenges the self-evidently progressive and productive nature of this understanding born of action and experience (a so-called engagement with the real). She argues that the cost of feminist success has been the widespread acceptance of a language of rape which

> solicits women to position ourselves as endangered, violable, and fearful and invites men to position themselves as legitimately violent and entitled to women's sexual services. This language structures physical actions and responses as well as words, and forms, for example, the would-be rapist's feelings of powerfulness and our commonplace sense of paralysis when threatened with rape. (Marcus 1992: 390)

More importantly for the argument being developed here, this "rape script" portrays women's bodies and female sexuality in spatial terms as an empty space waiting to be invaded/taken/formed:

> The rape script describes female bodies as vulnerable, violable, penetrable, and wounded; metaphors of rape as trespass and invasion retain this definition intact. The psychological corollary of this property metaphor

> characterizes female sexuality as inner space, rape as the invasion of this inner space, and anti-rape politics as a means to safeguard this inner space from contact with anything external to it. The entire female body comes to be symbolized by the vagina, itself conceived as a delicate, perhaps inevitably damaged and pained inner space. (Marcus 1992: 398)

This knowledge of woman's body space is not an artifact of purely philosophical reckoning but is a representation of the "reality" of women's bodies *vis-à-vis* men's. That this representation is informed by a movement from the concrete experience of rape victims and rapists to the abstract positioning of woman–space as absence/negativity and man–space as presence/positivity would attest to its legitimacy as true knowledge in Lefebvre's frame of reference.[9] Marcus's point, though, is that the language of rape is performative in the sense that it participates in constituting the condition it purports to describe. The rape script tends to defer and confine practical intervention to the postrape events of reporting, reparation and vindication, thereby blocking – or at least failing to encourage – an active strategy of rape prevention.[10] Thus much feminist knowledge of rape is bound by the language it employs to a perpetuation of victim status for women.

Marcus argues that the "truth" of victimhood should not be accepted but should continually be resisted and undermined. Her argument points up the problems with Lefebvre's view that "space (is) *produced* before being *read* . . . (it is produced) in order to be *lived* by people with bodies and lives in their own particular urban context" (1991: 143). According to Marcus, lived space is as much discursively as nondiscursively

[9] In Lefebvre's frame (one shared by many who do not identify with the poststructuralist camp), it is knowledge gained in and through an interaction between "reality" and "reflection" that affords "scientific understanding." This process of knowledge production is contrasted to that which involves analysis of texts/writing alone – a process which is destined, in Lefebvre's view, to reproduce an ideological understanding.

[10] That is, prevention beyond the legal deterrence of laws that are supposedly designed to persuade men not to rape or measures such as better street lighting which are designed to increase the public surveillance of male sexuality (Marcus 1992: 388). Klodawsky (1995) makes the point that the institutional forms taken by the anti-rape movement have tended to privilege service provision (in part because these projects can access government funding) over feminist projects addressed to changing the social conditions (including the socialization of men) that produce rape. It would seem that Marcus's argument is addressed to this tendency or imbalance within the movement against gendered violence but is specifically focused on the performative effects of a language of rape in the constitution of both female and male subjectivities: "The gendered grammar of violence predicates men as the subjects of violence and the operators of its tools, and predicates women as the objects of violence and the subjects of fear" (p. 393). It is this grammar that the focus on post-rape service provision or on increased surveillance and control over men fails to challenge.

produced. She urges us to produce a different discourse of female spatiality/sexuality, thus enabling a different female materiality/liveability.

A parallel construction of woman's body and female sexuality may be found in certain (feminist) knowledges of the city. Again, these knowledges are often based upon both the experience of women in the city and on contemporary theories of urban structure. From behavioral geographic research into gendered activity patterns and social networks a picture has been developed of women inhabiting certain spaces of the city – domestic space, neighborhood space, local commercial space, while men are more prevalent inhabitors of the central city, industrial zones and commercial areas. In urban studies women are often situated within the theoretical spaces of consumption, reproduction and the private, all of which are mapped onto the suburb (Wilson 1991, Saegert 1980, England 1991).[11] As vacuous spaces of desire that must be satisfied by consumption, women are positioned in one discourse as shoppers, legitimately entering the economic space of the city in order to be filled before returning to residential space where new and ultimately insatiable consumer desires will be aroused (Swanson 1995). As hallowed spaces of biological reproduction, women's bodies are represented in another urban discourse as empty, needful of protection in the residential cocoon where they wait, always ready to be filled by the function of motherhood (Saegert 1980).[12] Vacant and vulnerable, female sexuality is something to be guarded within the space of the home. Confined there, as passive guardians of the womb-like oasis that offers succor to active public (male) civilians, women are rightfully out of the public gaze (Marcus 1993).

In this type of urban theory the spatiality of women's bodies is constituted in relation to two different but perhaps connected Forms or Identities, that of the Phallus and that of Capital. These discourses of gender difference and capitalist development associate "woman" with lack, emptiness, ineffectiveness, the determined. As we have already seen in the rape script which is articulated within the broader hegemonic discourse of gender, woman is differentiated from man by her passivity, her vulnerability, ultimately her vacuousness. She is indeed the symbol of "absolute space," a homogeneous inert void, a container, something that

[11] Rose (1993) provides an excellent critical summary of the activity space literature. She notes that this literature, like much of feminist geography, draws largely upon the experience of white middle-class women in constructing the urban Woman, obscuring the very different geography of black urban women.

[12] In an intriguing reading of the film and novel *Rosemary's Baby*, Marcus (1993) alludes to the punishment that might befall any woman who deserts the fecundity and safety of the suburbs and the single-family house for the sterility and danger of the inner city and apartment living, yet proceeds to get pregnant and have a baby. (The devil takes such an out-of-place woman, or at least her child.)

can only be spoken of in terms of the object(s) that exist(s) within it.[13] Inevitably, the object that exists within/invades/penetrates the inert void – bringing woman into existence – is the Phallus. Woman is necessarily rape space in the phallocentric discourse of gender.

In the urban script which is articulated within the broader hegemonic discourse of Capitalism, woman is constituted as an economic actor allocated to the subordinate functions of the capitalist system. As consumer she is seen to participate in the realization of capitalist commodities, putting them to their final, unproductive uses; under the influence of capitalist advertising and mood manipulation she translates her sexual desires into needs which must be satisfied by consumption. This transfiguration of private into public desire is enacted in consumption spaces – the shopping mall, the high street, the department store – horizontal, sometimes cavernous, "feminized" places within the urban landscape. Represented as maker and socializer of the future capitalist workforce, woman plays a part in the dynamic of social reproduction. In her role of bearing children, ministering to their needs and assisting the state in their education and social training, woman is portrayed as an unpaid service worker attending to the requirements of capital accumulation. Within her limited field of action in the sphere of reproduction, resistance is possible – she may organize around local community and consumption issues – but the rules are made by Capital.

In this urban discourse woman is represented as an active player rather than a passive container; she is a crucial constituent of capitalist social relations, though not situated at the center of accumulation, nor cast as the subject of history.[14] The discourse of Capitalism renders the space of woman no longer homogeneous and void. Instead woman–space is "relative space," given form by multiple (subordinated) roles, each situated in relation to capitalist production. Women's economic bodies are portrayed as complements to men's economic bodies, adjuncts with important reproductive, nurturing and consumption functions. Indeed, woman becomes "positive negative space," a background that "itself is a positive element, of equal importance with all others" (Kern 1983: 152).

13 This is the Newtonian notion of space as a void, the "plenum of matter" (Kern 1983: 153; Smith and Katz 1993: 75).

14 This role is taken by man as the producer of commodities, the producer of surplus value, situated in the sphere of production, as a member of the working class. Of course recent episodes of industrial restructuring have altered the gendered face of the capitalist workforce. Women are increasingly occupying the sphere of production and the vertical concrete and glass spaces of economic power (McDowell 1994). However, a new urban discourse which dislodges the extremely gendered code that is mapped onto the suburb/sphere of reproduction/space of consumption is only beginning to emerge (Cameron 1995; Huxley 1995; Bell and Valentine 1995).

Like the structured backgrounds of cubist painting, woman–space as relative space is more visible, less empty, more functional than is absolute space.[15] But woman–space is still defined in terms of a positivity that is not its own. Whether as absolute or relative space, woman is presented as fixed by, or in relation to, an Identity/Form/Being – the Phallus or Capital.

In an attempt to address women's oppression, feminists may celebrate shopping, birth, homemaking, the fecund emptiness of woman's body, the shopping mall, the suburban home, the caring and nurturing functions, the woman–space. But in doing so they accept the boundaries of difference and separation designated by the discourses of capitalism and binary gender. Another feminist strategy has been to attempt to ignore or even reverse the spatialized binary by claiming back men's economic and urban space as rightfully women's. Women (particularly white female-headed households) have begun to desert the suburbs and, as one of the main groups involved in gentrification, have reasserted their right to a central location in the city (Rose 1989).[16] Women have successfully fought for child-care centers, vacation programs for school-age children, better community care for the elderly and disabled so that they can temporarily free themselves from the role of carer and claim a rightful place in the capitalist paid workforce (Fincher 1988). Indeed, the fact that such services are better provided for in cities contributes to the feminization of households in central urban areas. Significant though all these changes have been for women in the city, these strategies rest upon the assumption that women remain the carers, the supplementary workers in a capitalist system, who, if they undertake labor in the "productive" spheres of the economy must also provide the "reproductive" labor. The central city is one space that allows the (exhausted) middle-class superwoman to function – it has become the site of a new "problem that has no name."[17]

Similar strategies of reversal are represented in "Take back the night" rallies and other urban actions where women have claimed their right to the city streets, pressing for better lighting, better policing of public

[15] Kern (1983) argues that absolute space has more in common with the insignificant backgrounds of classical portraiture which serve only to contain and set off the foregrounded subject.

[16] Even in the face of foreboding and paranoid cautionary fables such as *Rosemary's Baby*.

[17] Marcus (1993) notes that after the publication of her bestseller, *The Feminine Mystique*, Betty Friedan was able to move out of the suburbs, the condemned site of women's unnamed oppression, and into an apartment in Manhattan. In the 1960s, unlike the 1990s, such an urban location represented an escape from the "problem that had no name."

transport, guarded parking stations, and other mechanisms of public surveillance of men's behavior (Worpole 1992). As the geography of women's fear has been made visible, so has the "reality" of male sexuality and the "inevitability" of violence against women been accepted. While greater public surveillance is advocated, women are simultaneously warned not to trespass into public space where, on the streets at night or on public transport after work hours, they are most certainly "asking" to become players in a rape script.[18]

Feminist strategies of celebration and reversal are all contributing to changes in the liveability of urban space for women. But what might be the cost of these changes if they rest upon the acceptance of both the Phallus and Capital as the "Identities" which define women/space, if they force women/space into the victim role that the sexual rape script allocates and the subordinate role that the economic urban script confers? What potentialities are suppressed by such a figuring of women and space? Perhaps we can only answer these questions by looking to alternative notions of Identity to see how they might differently configure women/space, as well as other possibilities they might entail.

Rethinking the space of Form: "air against earth"

Both absolute and relative conceptions of space rely upon the logic of Identity, presence or Form to give meaning to space. Absolute space is the emptiness which is the "plenum of matter" (Kern 1983: 153), "a passive arena, the setting for objects and their interactions" (Massey 1993: 76). Absolute space invokes a stable spatial ontology given by God, the Phallus, Capital, in which objects are fixed at an absolute location. Relative space comes into existence via the interrelations of objects (Massey 1993: 77). It invokes a fluid spatial ontology, continually under construction by the force fields established between objects. In Marxian formulations, all locations in absolute space are rendered relative by the dynamic historical structuring and restructuring of "capitalist patriarchy and racist imperialism" (Smith 1984: 82–3; Smith and Katz 1993: 79).

Not only is relative space historically and socially constructed, but space has its own effectivity:

> Could space be nothing more than the passive locus of social relations,

[18] In exposing the contradictions associated with this geography of fear, feminists have broken down the inside/outside distinction, citing the higher incidence of rape inside the home than outside it (Valentine 1992).

> the milieu in which their combination takes on body, or the aggregate of the procedures employed in their removal? The answer must be no. Later on I shall demonstrate the active – the operational or instrumental – role of space, as knowledge and action, in the existing mode of production. (Lefebvre 1991: 11)

Massey's development of Lefebvre's vision through the geological metaphor of sedimentation and layering has been influential in theorizing the effectivity of socially produced space (1984, 1993):

> no space disappears completely, or is utterly abolished in the course of the process of social development – not even the natural place where that process began. "Something" always survives or endures – "something" that is not a *thing*. Each such material underpinning has a form, a function, a structure – properties that are necessary but not sufficient to define it. (Lefebvre 1991: 403)

> What takes place is the interrelation of the new spatial structure with the accumulated results of the old. The "combination" of layers, in other words, really does mean combination, with each side of the process affecting the other. (Massey 1984: 121)

What is interesting in all these spatial conceptions is the prevalence of the image of space as ground or earth (Lefebvre's "material underpinning") – something which gives the ahistorical Identity located in absolute space a "place to stand" or the historically grounded Identity of relative space a "terrain" to (re)mold. But what effect does this reliance upon Identity and the metaphor of grounding have? What violence might it do to space?

Amongst other poststructuralist theorists who challenge the metaphysics of presence in western post-Enlightenment thought, Deleuze and Guattari employ a spatiality that appears divorced from the positive form of Identity. Rather than positing earth, ground, and fixity in a locational grid, their space evokes air, smoothness and openness:

> The space of nomad thought is qualitatively different from State space. Air against earth. State space is "striated" or gridded. Movement in it is confined as by gravity to a horizontal plane, and limited by the order of that plane to preset paths between fixed and identifiable points. Nomad space is "smooth," or open-ended. One can rise up at any point and move to any other. Its mode of distribution is the *nomos*: arraying oneself in an open space (hold the street), as opposed to the logos of entrenching oneself in a closed space (hold the fort). (Massumi 1987: xiii)

In the wild productions of "rhizome" thought, Deleuze and Guattari splinter Identity into disorder, chaos, multiplicity, heterogeneity, rupture, and flight. It is mapped rather than traced: "The map is open and connectable in all of its dimensions; it is detachable, reversible, adapted to any kind of mounting, reworked by an individual, group, or social formation" (Deleuze and Guattari 1987: 12). And mapping, as Carter has argued, is not about location and discovery (of already established identity) but about exploration and invention: "To be an explorer was to inhabit a world of potential objects with which one carried on an imaginary dialogue" (1987: 25).[19]

These images of space as air and openness, enabling exploration and liberating potentiality, evoke feminist and postmodern uses of chora to represent space (Grosz 1995; Lechte 1995):[20]

> *chora* is fundamentally a space. But it is neither the space of "phenomenological intuition" nor the space of Euclidean geometry, being closer to the deformations of topological space. Indeed, the *chora* is prior to the order and regulation such notions of space imply. It is an unordered space. Although Kristeva herself says that the *chora* "preceded" nomination and figuration, this is not meant in any chronological sense. For the *chora* is also "prior" to the ordering of chronological time. The *chora*, therefore, is not an origin, nor is it in any sense a cause which would produce predictable effects. Just the reverse: the *chora*, as indeterminacy, is a harbinger of pure chance. (Lechte 1995: 100)

Chora is the term Plato uses to denote the space of movement between being and becoming – "the mother of all things and yet without ontological status":

> Chora then is the space in which place is made possible, the chasm for the passage of spaceless Form into a spatialized reality, a dimensionless tunnel opening itself to spatialization, obliterating itself to make others possible and actual. (Grosz 1995: 51)

The femininity of chora lies in its immanent productiveness. But it is this very quality that Grosz argues has been undermined by phallogocentrism.

[19] Others, particularly postcolonial analysts, have convincingly demonstrated the ways in which mapping has been used as a graphic tool of colonization and imperial power (Blunt and Rose 1994; Harley 1988).

[20] Having assumed the status of a "master term" within French poststructuralist thought, *chora* is of interest because of the way in which it cannot be contained within the logos of any text under examination but is, nevertheless, necessary to the operations of that text.

For Derrida and Kristeva such a term highlights the limits or excess of a system of thought, the vulnerable point at which to focus the most productive deconstruction (Grosz 1995: 48).

Within phallocentric thought chora became appropriated as the space of Form/the Father/Production – the space which is the condition of man's self-representation and the condition of Identity. Chora as the space of indeterminacy/enabling/engendering/the mother was denigrated, represented "as an abyss, as unfathomable, lacking, enigmatic, veiled, seductive, voracious, dangerous and disruptive" (Grosz 1995: 57), cast out without name or place.

Feminist poststructuralists have been keen to point out the violence that has been done to women, and now to space, by phallocentric modernist discourse:

> (The) enclosure of women in men's physical space is not entirely different from the containment of women in men's conceptual universe either: theory, in the terms in which we know it today, is also the consequence of a refusal to acknowledge that other perspectives, other modes of reason, other modes of construction and constitution are possible. Its singularity and status, as true and objective, depend on this disavowal. (Grosz 1995: 56)

How might we proceed now to reclaim the feminine aspect of chora, to conceptualize a pregnant space, a space of air, a space of potentiality and overdetermination?

> In order for [sexual] difference to be thought and lived, we have to reconsider the whole problematic of *space* and *time* . . . A change of epoch requires a mutation in the perception and conception of *space–time*, the *inhabitation of place* and the *envelopes of identity*. (Irigaray, quoted in Grosz 1995: 55)

> Becomings belong to geography, they are orientations, directions, entries and exits. (Deleuze and Parnet 1987: 3)

Feminist theorists urge us to think woman and space outside of that discourse in which Identity, or the Phallus, gives meaning to everything – to think outside the discourse in which woman can only be given shape by Man and in which space is an empty container that can only be given shape by matter.[21] To this urging can be

[21] ". . . identities based on spatial containment, substances and atoms belong to the *masculine imagery*, and what is missing from our culture is an alternative tradition of thinking identity that is based on fluidity or flow. It is important to note that Irigaray is not making an experiential claim: she is not asserting that women's true identity would be expressed in metaphors and images of flow. What she is claiming, by contrast, is that identity as understood in the history of Western philosophy since Plato has been constructed on a model that privileges optics, straight lines, self-contained unity and solids . . . the Western tradition has left unsymbolised a self that exists as self not by repulsion/exclusion of the not-self, but via interpenetration of self with otherness" (Battersby [1993: 34] on Irigaray).

added the encouragement, offered by anti-essentialist Marxism, to think economy and space outside the discourse in which Identity, or Capital, is the origin of social structure and intelligibility – to think outside the discourse of woman's economic subordination to Man and of urban women operating in a terrain defined by capitalist social relations.

Pregnant space, postmodern space

Geographic and feminist projects of representing space find a number of parallels within the visual arts. Impressionism and Cubism, for example, are two interrelated art movements which mirror the possibilities and potentialities of, as well as the impossibilities and barriers to, thinking a postmodern pregnant space. In the paintings of the Impressionists space was, for the first time, constitutive – the background, full of haze, mist, smoke, light, crowded in on the subject, claiming equal status and attention from the painter and gazer (Kern 1983: 160). Cubism took one step further, instating space with geometric form, leveling space and material object to the point of complete interpenetration. In this genre Form was both disintegrated and reFormed in every constituent space (Kern 1983: 161–2).

Cubism, however, evokes a closed system of determination in which space is defined by the presence of a positive Being, no matter how fragmented and indistinct. Cubist painting can thus be seen to represent the quintessential space of modernism, paralleling in the visual realm the discursive space of phallocentrism and the economic space of capitalism. In the discourse of hegemonic Capitalism, for example, all space is constituted by the operations of capital:

> It is not Einstein, nor physics and philosophy, which in the end determine the relativity of geographical space, but the actual process of capital accumulation. (Smith 1984: 82–3)

> The new space that thereby emerges [in the moment of the multinational network, or what Mandel calls "late capitalism"] involves the suppression of distance . . . and the relentless saturation of any remaining voids and empty places, to the point where the postmodern body . . . is now exposed to a perceptual barrage of immediacy from which all sheltering layers and intervening mediations have been removed. (Jameson 1991: 412–13)

> It is here that we can begin to see the relation between capitalism and the construction of everyday life as a transit-mobility which constructs the

> space for the free movement of capital and for the capitalization, rather than the commodification, of everyday life. For within that transit-space, people are not the producers of wealth but a potential site of capital investment. People become capital itself. And within these circuits, the only thing that they can be sure of is that capitalism is going first-class. (Grossberg 1992: 328)

Although "we gotta get out of this place" (Grossberg 1992) we are caught in a space of no escape.

By contrast, the space of the Impressionists could be seen to represent one of those points of excess within modernism. Amidst modernist and realist attempts to replicate the fracturing of light to heighten the experience of color, Impressionists such as Monet and Pissaro painted evanescent atmospheric effects (reminiscent of Turner, a British forerunner) in which form and order are "destroyed." Space is constituted by the random distribution, disorder and chance of smoke, streams of sunlight, steam and clouds (Lechte 1995: 101). Here we see space represented as an open system of disequilibrium and indeterminacy, a random but productive *process* (Serres, cited in Lechte 1995). In this chora-like image of positive immanence and potentiality it might be possible to see postmodern becomings that are not devoid of political in/content.

How might we, for example, appreciate differently the spatiality of female sexuality and potential new ways for women to dwell in urban space? Marcus provides some guidance:

> One possible alternative to figuring female sexuality as a fixed spatial unit is to imagine sexuality in terms of time and change . . . Rather than secure the right to alienate and own a spatialized sexuality, antirape politics can claim women's right to a self that could differ from itself over time without then having to surrender its effective existence as self. (Marcus 1992: 399–400)

Marcus appears to be arguing for a multiplicity of female sexualities that may coexist within any one individual. In her vision, the spatiality of female sexuality can be dissociated from the notion of a fixed, immobilized cavity defined in relation to the inevitably invading, violent penis. Instead, female sexual space can be conceived in multiple ways – as surface, as active, as full and changing, as many, as depth, as random and indeterminate, as process.[22]

How might this respatialization of the body contribute to new geographies for women in the city? It might lead us to identify the

[22] Marcus goes on, in fact, to rewrite the rape script in the light of this conception, and I draw upon her "revised rape script" in chapter 6. In this chapter, it is her rethinking of the spatiality of the body that I am interested in.

multiple urban spaces that women claim, but not solely in the name of consumer desire or reproductive/biological function. Here one could think of the heterotopias of lesbian space, prostitution space, bingo space, club space, health spa, body building and aerobics space, nursing home space, hobby space – all terrains of public life in which women's agency is enacted in an effective, if indeterminate manner.[23] One could identify the ways in which such spaces are regulated and ordered by dominant discourses of heterosexuality, health, youth, beauty, and respectability and influenced by discourses of transgression.[24] One could explore and map an urban performance space of women that is defined in terms of positivity, fullness, surface and power. But in order for such a reinscription not to fall back into simply celebrating woman–space in the city,[25] theoretical work must continually and repeatedly displace (rather than only reverse) the binary hierarchy of gender.

One strategy of displacement might lead us to deconstruct and redefine those consumption and reproductive spaces/spheres that are the designated woman–space in the discourse of urban capitalism. Within geography, for example, the urban restructuring literature points to the massive involvement of women in the paid workforce where they are active in a variety of economic roles apart from that of final consumer or reproducer of the capitalist labor force. Feminist geographers and sociologists are researching women in office space (Pringle 1988), in finance space (McDowell 1994), in retail space (Dowling 1993), in ethnic small business (Alcorso 1993), in industrial space (Phizacklea 1990) – again all public arenas in which women's

[23] Some of this work is currently being done by feminist geographers. Many of the early studies in feminist geography have, however, reproduced a phallocentric discourse by accepting the representation of women's bodies as vulnerable and women's spaces as subordinate (see Rose 1993: 117–36).

[24] In a project which "reflects the intense realism underpinning any queer utopian impulse" (p. 30), Moon et al. (1994) detach the suburban house from its pre-eminent representation as a container for heterosexual couples and their families. Redefining the house as a (not exclusively but nevertheless) queer space, where all manner of sexual practices and relations are enacted, they simultaneously redefine space as something that cannot be definitively dedicated to particular activities or exhaustively structured by a single form or "identity," such as the heterosexual family: "Queer lives and impulses do not occupy a separate social or physical space from straight ones; instead, they are relational and conditional, moving across and transforming the conventional spaces that were designed to offer endless narcissistic self-confirmation to the unstable normative systems of sex, gender and family" (p. 30). This space is open, full of overlaps and inconsistencies, a place of aleatory relations and redefinitions, never fully colonized by the pretensions of a singular identity.

[25] As Soja and Hooper (1993: 198) suggest, the task is not to "assert the dominance of the subaltern over the hegemon" but to "break down and disorder the binary itself."

agency is enacted.[26] In some texts we may even see glimmers of spaces beyond or outside capitalism, where women operate in noncapitalist spaces of production and contribute to the reproduction of noncapitalist economic forms.[27]

Despite these glimmers, what characterizes much of the restructuring literature is an overriding sense of "capitalocentrism" in that women's entry into the paid labor force is understood largely in terms of the procurement by capital of cheaper, more manipulable labor. Capital has positioned the superexploited female worker just as it has produced women's roles as reproducers (of the capitalist workforce) and consumers (of capitalist commodities). Any attempt to destabilize woman's position and spatiality within urban discourse must dispense with the Identity of Capitalism as the ultimate container[28] and constituter of women's social and economic life/space.

It would seem that the rethinking of female sexuality and the creation of alternate discourses of sexuality and bodily spatiality are well in advance of the rethinking of economic identity and social spatiality (Grosz 1994b). Indeed, even the most innovative cultural and poststructuralist theorists tend to leave this terrain untouched:

> Individuals do not appear to appropriate capital but to be appropriated to it. People are caught in its circuits, moving in and out of its paths of mobility, seeking opportunistic moments (luck, fate, fame or crime) which will enable them, not to redistribute wealth, but to relocate themselves within the distributional networks of capital. (Grossberg 1992: 328)

[26] In a related move Staeheli (1994) attempts to break down the public/private binary that often underpins a vision of women as largely excluded from the public sphere. Arguing that the boundary between public and private is "fuzzy" and always being (re)constructed, she dissociates public acts from public spaces and public identities, dissolving the notion of a public (political) sphere in which they all come together, and hoping thereby to liberate the transgressive political potential of public acts in private spaces (e.g., home-based political organizing involving neighbors and children) and private acts in public ones (e.g., breast-feeding in restaurants). In a similarly disruptive piece entitled "Semipublics" Moon (1995) explores the very public nature of what is ostensibly private, arguing that "not just some but *all* sexualities in our culture are phantasmatically staged in public." Whereas the "audience-orientation and public-directedness" of the "bourgeois conjugal bedroom" is nominally associated with privacy, very private acts of transgressive sexuality are "relegated to the scandalous realm of 'sex in public'" (pp. 2–3).

[27] See, for example, Katz and Monk (1993) and chapter 10 below.

[28] So that household labor and self-employment (which may be understood as outside capitalist relations of production) are seen as somehow taking place "within capitalism," as are noncapitalist forms of commodity production (e.g., independent or communal production).

> The capitalist relation consists of four dense points – commodity/consumer, worker/capitalist – which in neoconservative society are effectively superposed in every body in every spacetime coordinate. When capital comes out, it surfaces as a fractal attractor whose operational arena is immediately coextensive with the social field. (Massumi 1993: 132)

Despite the postmodern interest in chora, in nomadology and smooth space, the identity of Capital confronts us wherever we turn. Do we only ever dwell in a capitalist space? Can we ever think outside the capitalist axiomatic?

> The economy constitutes a worldwide axiomatic, a "universal cosmopolitan energy which overflows every restriction and bond," (Marx) a mobile and convertible substance "such as the total value of annual production." Today we can depict an enormous, so-called stateless, monetary mass that circulates through foreign exchange and across borders, eluding control by the States, forming a multinational ecumenical organization, constituting a supranational power untouched by government decisions. (Deleuze and Guattari 1987: 453)

Here Deleuze and Guattari are difficult and elusive. Their capitalist axiomatic is all-pervasive and innovative, seemingly able to coopt and reterritorialize all lines of flight out of its territory into new opportunities for self-expansion, able to set and repel its own limits (1987: 472). Yet at the same time they reserve a space for the minority, for the becoming of everybody/everything outside the totalizing flow of capital:

> The undecidable is the germ and locus par excellence of revolutionary decisions. Some people invoke the high technology of the world system of enslavement: but even, and especially, this machinic enslavement abounds in undecidable propositions and movements that, far from belonging to a domain of knowledge reserved for sworn specialists, provides so many weapons for the becoming of everybody/everything, becoming-radio, becoming-electronic, becoming-molecular . . . 68 Every struggle is a function of all these undecidable propositions and constructs revolutionary connections in opposition to the conjugations of the axiomatic. (1987: 473)

In the footnote to this statement the authors mention the domain of "alternative practices" such as pirate radio stations, urban community networks, and alternatives to psychiatry (1987: 572). Here we catch a minimal glimpse of what might lie outside the flows of Capital. The capitalist axiomatic closes and defines – in the sense of fully inhabiting – social space (evoking the closure and definition of Cubism), yet it is also in motion, providing a space of becoming, of undecidability. This space is reminiscent of the constitutive (pregnant) space of the Impressionists. It is a space of mists and vapors, of movement and possibility, of background that might at any moment become foreground – a "space of excess" and indeterminacy within the modern space of fullness and closure.[29]

If we are to take postmodern spatial becomings seriously then it would seem that we must claim chora, that space between the Being of present Capitalism and the Becoming of future capitalisms, as the place for the indeterminate potentiality of noncapitalisms.[30] In this space we might identify the range of economic practices that are not subsumed to capital flows.[31] We might see the sphere of (capitalist) reproduction as the space of noncapitalist class processes that deterritorialize and divert capitalist flows of surplus value (see chapter 9). We might see the sphere of (capitalist) consumption as the space of realization and consumption of commodities produced under a range of productive relations – cooperative, self-

[29] Negri offers images of capitalism expanding to encompass and cover every social and cultural domain but stretching so much in the process that it begins to thin and tear, creating openings for resistance and "islands of communism" (1996: 66).

[30] Here we may enter a space resembling Bhabha's third space "beyond the discursive limits of the master subject" (Blunt and Rose 1994), or the "thirdspace of political choice" depicted by Soja and Hooper (1993: 198–9) (drawing on Foucault's [1986] notion of heterotopia) which is a place of enunciation of a "new cultural politics of difference." Such a space also resembles Rose's (1993) "paradoxical space," a space that is productive of multiple and contradictory identities, or that of de Lauretis who discovers in the "elsewhere" or "space-offs" of hegemonic discourses the interstices in which the "subject of feminism" may emerge (quoted in Rose 1993: 139–40).

[31] See Arvidson (1996) for an attempt to theorize urban development in Los Angeles outside a vision of hegemonic capitalism. While theorists of capitalism have come to acknowledge that capitalist spaces are "coinhabited" by noneconomic relations (including racism, sexism, heterosexism, and so on), the space-economy itself is most commonly represented according to what Rose calls the "masculinist" principles of exhaustiveness and mutual exclusivity. Thus capitalism generally covers the entire social space (see Massumi above, for example, where capital is "coextensive with the social field") and tends not to coexist with noncapitalism in the same location. To undermine the closures and exclusions of these colonizing representations, Rose calls at one point upon the work of Mackenzie (1989a), which questions the neat spatial and social "dichotomies of the divided city" (p. 114, cited in Rose 1993: 135). Mackenzie's work on women's labor in informal networks destabilizes the familiar division of the city into spaces of (capitalist) production and reproduction, discovering activities which are neither, taking place in spaces usually identified with one or the other. So unusual is it for economic representations to be set outside the imperial space of the master term that Mackenzie is at a loss to name and conceptualize these noncapitalist activities. She prefers to let them go "conceptually unclad . . . so to speak" (1989b: 56).

employed, enslaved, communal as well as capitalist.[32] What violence do we do when we interpret all these spaces as existing *in* Capitalism, as cohering within the coded flows of axiomatic capital? We risk relegating space/life to emptiness, to rape, to non-becoming, to victimhood.

[32] Daly offers a similar vision of a rich and pregnant economic space: "Economic identity must always be regarded as provisional and contingent. This is why I want to talk about the economic as a space rather than a model: not a given space, but a space of possibilities dominated by a proliferation of discourses which are always capable of subverting and rearticulating the identities that exist there . . . It is clear . . . that a whole range of radical enterprises exist within the sphere of the market, including: credit unions, co-operatives of every type, housing associations, radical journals/literature, alternative technology, alternative forms of entertainment, etc., as a counter-enterprise culture; none of which can be regarded as having an unequivocal status as "capitalist" (1991: 88).

5

The Economy, Stupid![1] Industrial Policy Discourse and the Body Economic

> Once upon a time, people used to talk about ISSUES and HAVE FUN. But then someone invented the economy . . . The economy grew and grew! It took over EVERYTHING and NO-ONE COULD ESCAPE.
>
> (Morris 1992: 53, quoting from memory a recent cartoon)

> I saw men on television (trade-union stars, Cabinet Ministers, left-wing think-tank advisers) visibly hystericized by talking economics: eyes would glaze, shoulders hunch, lips tremble in a sensual paroxysm of "letting the market decide," "making the hard decisions," "leveling the playing field," "reforming management practices," "improving productivity" . . . those who queried the wisdom of floating the exchange rate, deregulating the banks, or phasing out industry protection were less ignored than washed away in the intoxicating rush of "living in a competitive world" and "joining the global economy."
>
> (Morris 1992: 51–2)

In *Ecstasy and Economics*, Meaghan Morris chronicles the ecstatic submission of white Australian men to "the economy."[2] Humbled before its godlike figure, grown men grovel and shout in fundamentalist rapture, transported in "an ecstasy of Reason" (1992: 77). By giving themselves over to a higher power, they have paradoxically gained mastery and authority. They "talk economics" and find themselves speaking the

[1] A sign allegedly posted in Clinton headquarters to remind campaign workers of the central issue of the 1992 presidential campaign.

[2] As Fred Block points out, the economy has increasingly become the social site which dictates or constrains social policy: "a broad range of social policies are now debated almost entirely in terms of how they fit in with the imperatives of the market" (1990: 3).

language of pure necessity, unhampered by base specificities of politics and intention. In the face of necessity, and in its despite, they project a wilful certainty that their economic "interventions" will yield the outcomes they desire.

During the 1980s and 1990s Australia has been one of the few OECD countries governed by a social democratic (albeit right-wing) Labor Party in which interventionist economic and industrial policies have been on the national agenda. Recently, though abortively, the Clinton administration promised to concern itself with many of the things that concerned the Hawke and Keating governments from the beginning: deindustrialization, lack of technological innovation, a labor force unsuited to the needs of industry, a weak competitive position in a rapidly changing world. In seeking models of successful intervention that have presumably fostered rather than blocked economic adaptation, American economic strategists looked to Australia for innovative ways of meeting Clinton's mandate to "grow the economy." These American analysts included not merely center and right-wing Democrats but Marxists and other leftists whose pronouncements were suddenly contiguous to debates in the mainstream press.

After 12 years (or maybe a lifetime) in exile, leftists in the US were "talking economics" in a room where just possibly they could be overheard. And the economics they were talking was in some ways very different from what was permissible just a few years before, when "industrial policy" or "managed trade," for example, could not be broached at the national level. Yet despite its release from old strictures and prohibitions, the discussion of economic policy seemed entirely familiar. It moved laboriously in a confined space, as though hobbled by an invisible tether or circumscribed by a jealous and restrictive force – something more potent even than the political realities that also operate to keep debate within narrow and familiar limits.

Despite their divergent positions on every issue, the right and left share a "discourse of economy" that participates in defining what can and cannot be proposed. What from a right-wing perspective may seem like a truly misguided left-wing proposal is nonetheless intelligible and recognizable as a member of the extended family of potential economic initiatives, and vice versa. This is not to say that right- and left-wing policy analysts profess the same economic theories and harbor the same social conceptions. In their positive proposals, their understandings of economy and society are often revealed to be quite different, and indeed they may have been trained in very different schools of thought.[3] Nevertheless, there seems to be a substrate of commonality, detectable in the ubiquitous affective paradoxes of submission and control, arrogance and caution, that structure the range of economic emotions. If the economy of the left is so different in its operations and possibilities from that of the right, why does it produce such similar affective disjunctions? Why is "the economy" at once the scene of abject submission, the social site that constrains activities at all other sites, the supreme being whose dictates must unquestioningly be obeyed and, at the same time, an entity that is subject to our full understanding and consequent manipulation? And how is it, furthermore, that something we can fully understand and thus by implication fully control is susceptible only to the most minimal adjustments, interventions of the most prosaic and subservient sort? What accounts for the twin dispositions of utter submission and confident mastery, and for boldness and arrogance devolving to lackluster economic interventions?

Of course, these questions could be turned upon the questioner, and one might wish to understand how it is that I am positioned to see the left and the right as operating within the same "discourse of economy" despite the cacophony produced by their different starting places, their divergent ends and means, their backgrounds in Marxism or neoclassicism, their heterogeneous present attachments to Keynesianism, post-Keynesianism, and various forms of development economics. In what discursive space am I situated, that left proposals appear strangled and truncated rather than as reasonable or even as exhausting the realm of the possible? If I turned to cultivating that space, to "growing an alternative discourse of the economy," what monstrous novelties might emerge?[4]

The task of cultivation is so daunting that I scarcely know where to begin. But fortunately I do not have to make a beginning, since I too ambpart of a lineage. Indeed, I can only locate myself outside the "discourse of the economy" by virtue of my association with an alternative economic knowledge, even though the products of that knowledge are few and far between.[5] What follows, then, can be read as the delineation of an existing formation whose magnificent contours can suddenly be seen from the vantage of a new and separate space, itself uncultivated and unformed.

[3] Nor is it to suggest that leftists (or for that matter right-wingers) are unified in their economic thinking; or to deny that very different policy proposals will produce very different economies, belying the notion of a singular "economy" or economic conception.

[4] Haraway (1991) asks a similar question as she embarks on her monstrous project of "reinventing nature."

The body economic

Ailments in search of a cure

Anorexia, meaning without appetite, is a starvation syndrome that has reached epidemic proportions in wealthy western social formations. Deindustrialization, defined as the decline of traditional manufacturing, is an economic condition widely perceived as a threat to the industrial capitalist nations. What might be the connection between these two representations of disorder?

A solution to this riddle can be found in the ways in which medical interventions into anorexia, and industrial policy interventions into deindustrialization, are construed as potential "cures" for the ailments of a suffering body. Food is administered intravenously to the anorectic, and investment is lured to declining industrial regions, in order to revitalize an ailing corporeal being. Convincing the anorectic to participate in family therapy and negotiating with the downsized workforce to stem wages growth and introduce a new work culture are both attempts to foster the conditions under which the essential life forces, calories and capital, might restore the body to its natural state of health.[6]

Twenty years of investment policies directed at declining industries and regions have resulted in only marginal success in redressing the deindustrialization disorder. Yet there are few attempts to rethink the economic discourse upon which this "cure" is predicated. By contrast, the human body is currently the focus of a radical rethinking (see, for example, Bordo 1989; Gatens 1991; Grosz 1994b; Kirby 1992). Feminists exploring the social construction of the female body have questioned the centrality of the phallus, or its lack, in governing the actions of the embodied subject. The body is reappearing as a fluid, permeable and decentered totality in which physiological, erotic, mental, psychological, social and other processes mutually constitute each other,

[5] They include the emerging postdevelopment discourse exemplified in the work of Arturo Escobar (e.g., 1995) and others; various attempts to "marginalize the economy" in order to re-vision the conditions of social possibility (e.g., Block 1990); and the journal *Rethinking Marxism*, which is a site of the reinvention of Marxism as a discourse of overdetermination and anti-economistic social analysis (see as well Resnick and Wolff 1987).

[6] In a fascinating dialogue around the complicated association of female fatness with economic accumulation and waste, Michael Moon and Eve Kosofsky Sedgwick discuss the emergence of a Dickensian loathing and revulsion toward the fleshy female body in post-Enlightenment Western culture. They point to the shift, after World War I, "of thinness from being a lower-class to an upper-class female signifier" and to the delicate negotiation between representations of overeating as "unhealthy" and excessive dieting as "addiction" within the medicalized discourse of fat (1993: 233–4).

with no one process or zone being more invested with meaning or effectivity than another.

In part what has motivated this rethinking are the social effects of representing the (female) body as a bounded and structured totality governed by the psyche (or some other locus of dominance) instead of as a "material-semiotic generative node" with boundaries that "materialize in social interaction" (Haraway 1991: 200–1). The physical and psychological tortures associated with the treatment of anorexia, for example, have prompted a reconceptualization of the body as a complexly overdetermined social site rather than a discrete entity subject to internal governance and medically restorable to self-regulation. Thus psychotherapist Harriet Fraad sees anorexia as an agonized crystallization of the contradictions "crowding in on [women's] lives" (1994: 131) as men, bosses, the media and women themselves exercise new and demanding expectations of women.

For Fraad, the body is both a site where the female subject takes control and resists social, sexual and economic expectations, and a site where control is relinquished as the anorectic takes to heart the body image associated with "success" as an object–woman.[7] The body is an overdetermined social location in which a multitude of social, political, physiological, and discursive practices participate in constituting the act of starvation. From the standpoint of this representation, the medical and psychological treatment of anorexia that focuses upon the individual and her family is addressing only a very few of the contradictory practices constituting the anorectic condition, and therefore has only limited potential as a cure.

Whereas feminist theorists have scrutinized and often dispensed with the understanding of the body as a bounded and hierarchically structured totality, most speakers of "economics" do not problematize the nature of the discursive entity with which they are engaged. Instead, they tend to appropriate unproblematically an object of knowledge and to be constructed thereby as its discursive subjects. In familiar but paradoxical ways, their subjectivity is constituted by the economy which is their object: they must obey it, yet it is subject to their control; they can fully understand it and, indeed, capture its dynamics in theories and models, yet they may adjust it only in minimal ways. These experiential constants of "the economy" delineate our subjective relation to its familiar and unproblematic being.

[7] Grosz argues that anorexia is "a form of protest at the social meaning of the female body. Rather than seeing it simply as an out-of-control compliance with the current patriarchal ideals of slenderness, it is precisely a renunciation of these 'ideals'" (1994b: 40). I would argue, with Fraad, that it could be both.

Constituted in relation to the economy as both submissive and manipulative beings, capable of full knowledge but of limited action, our political effectivity is both undermined and overstated. With the consummate and ultimately crippling arrogance of modernist humanism, we construct ourselves as both the masters and the captives of a world whose truth we fully apprehend. In the face of that world or, more specifically, of the discourse of its economic form, and in the trains of the subjectivity which that discourse posits and promotes, we struggle to mark the existence and possibility of alternate worlds and to liberate the alternative subjectivities they might permit. But in order to recreate or reinform the political subject – a project which is arguably a rallying point for left social theory in the late twentieth century – it is necessary to rethink the economic object. Given the centrality of the economy to modernist social representations, and given its role in defining the capacities and possibilities of the left, it is necessary to defamiliarize the economy as feminists have denaturalized the body, as one step toward generating alternative social conceptions and allowing new political subjectivities to be born.

The birth of the organism: metaphors of totality and economy[8]

Like the anorectic woman constructed as a target of medical intervention, the economy of the economic strategists and planners is depicted as a body, and not just any body. It is a bounded totality made up of hierarchically ordered parts and energized by an immanent life force.

[8] The movement among some economists to view economics as discourse, that is, as a site in which meanings are continually negotiated and ultimately unfixed, has generated a growing interest in metaphor among economic discourse analysts, who range from the relatively apolitical to the explicitly political in their interests and intentions. For McCloskey (1985), metaphor is but one of the devices used in the contest of rhetoric between competing paradigms. Thus, for example, the appropriation of physics metaphors by neoclassical economists was an attempt to establish scientific status for their emerging paradigm (Mirowski 1987: 159); it was part of a disciplinary process of self-justification, involving a quest for the appearance of rigor as well as ontological validation for privileging the individual within a theory of society.

In contrast to most though not all economic discourse analysts (see, for example, the work of Jack Amariglio and Antonio Callari), Foucault and Haraway are interested in metaphors for their social and political effects. Foucault is concerned with the ways in which power and knowledge intersect within economic discourse to enable particular conceptions of acting subjects and, within the modern episteme, to participate in producing Man (Amariglio 1988: 609). Haraway is motivated to deconstruct the metaphors through which we have understood society (both human and animal) in order to foster liberation and the building of "new relations with the world" (1991: 19).

In a word, the body economic is an *organism*, a modern paradigm of totality that is quite ubiquitous and familiar.

The organismic totality emerged, by some accounts, with the birth of "the economy" as a discrete social location.[9] When Adam Smith theorized the social division of labor as the most productive route to social reproduction, he laid the groundwork for a conception of "the economy" as a coherent and self-regulating whole (Callari 1983: 15).[10] By analogy with the individual who labored to produce his own means of subsistence, thereby constituting a unity of production and consumption, Smith saw society as structured by a division of labor among quintessentially "economic" human beings laboring for their own good and achieving the common good in a process of harmonious reproduction.

[9] Haraway (1991: 7) argues that, at the beginning of the industrial revolution in Europe, the representation of both nature (the natural economy) and political economy in terms of the body resuscitated organic images of the body politic developed by the ancient Greeks. While it is usually thought that economics in particular and social science in general poached their metaphors from physics and biology, actually economics has provided the source for some of the most well known metaphors of the natural world – including that of the organism and the metaphors employed in understanding evolution. Perhaps the most famous instance is Darwin's story of the way in which his own narrative of competition and struggle was inspired by the writings of Smith and Malthus.

Mary Poovey describes the emergence in eighteenth-century England of "the economy" as a distinct and bounded social domain in terms of a discursive object embedded in and giving shape to other aspects of social life: "The term *economy* initially referred to the management of a household . . . In the course of the eighteenth century, the word *economy* was yoked to the term *political* and used to signal the management of national resources . . . the economic domain can be seen as an Imaginary entity that is governed by a specific rationality, in this case, the logic and procedures by which productivity and financial security are thought to be ensured . . . Institutionally, the rudiments of what eventually became the economic domain were established in England in the late seventeenth century, in the Bank of England, the national debt, and the stock exchange. These institutions, in turn, along with the discipline by which they were detailed and naturalized – political economy – constituted the first of many concrete forms in which individuals encountered and imagined *the economic* to exist" (1994: 8–9).

[10] According to Callari (1983), Smith's theoretical object was to conform the homogeneity of human interests (the universal need for survival) with the heterogeneity of class positions (differential positioning with relation to the means of survival) that characterized a capitalist social formation. His was a quintessentially political project – to justify capitalism and its inequalities, including the existence of a class of propertyless individuals, within a social context in which an equalizing doctrine of needs and rights common to all men had been articulated and would prevail. By framing society as a unity in which inequalities of property and class were both a requisite and a guarantor of greater social well-being, Smith not only achieved his political objectives but set the stage for the emergence of "the economy" as a bounded and unified social instance.

In the absence of specialization producers are atomized, producing on their own or in small communities the wealth that satisfies their wants and needs; the "economy" is a plurality of practices scattered over a landscape. Increased specialization, however, requires greater social integration, in order for reproduction to take place. The *division of labor*, and the specialization it entails, thus necessitates the *integration of labor*.[11] Over the course of history, then, what was once plural becomes singular. Fragmentation becomes an aspect of unification rather than a state of atomism and dispersal. Scattered economic practices come together as "the economy" – something we all recognize, though may differently define, in economic discourse today.

Eighteenth-century students of animal nature adopted the vision of "the social economy" as a metaphor for the animal body, even referring to the latter as an "animal economy," which they envisioned as "various organ parts or functions" operating in a coordinated "division of labor" for the common good (Canguilhem 1988: 88). Drawing on the developing lore of machinery, these founders of modern physiology used the notion of an internal regulator or governor[12] to understand the way in which "organ systems seemed to be controlled from within" (p. 88) and had the capacity to maintain an equilibrium or "normal" state.

A developing vitalism breathed life into these conceptions, ascribing to human and animal bodies "some inherent power of restitution or reintegration" (p. 89). "Life" makes the organism susceptible to death and disease but also gives it the capacity for recovery (p. 132), the ability to re-establish wholeness or "health" in accordance with its telos or life form (p. 129). As the organism's invisible sovereign, "life" connects the internal to the external, the visible to the invisible, producing the "coherent totality of an organic structure" (Foucault 1973: 229). Its presence establishes reproduction of the organism (the struggle against death) as its raison d'être.

It is relatively easy to read certain forms of Marxian theory as tracing the lineaments of an economic body. In many versions of Marxism, the capitalist economy or society is represented as a totality governed and

11 See Sayer and Walker (1992). Buck-Morss (1995: 449) points to the paradox inherent in this otherwise elegant vision – the real bodies of workers become stunted and stultified by the nature of the divided labor they are required to do "in order for the social body to prosper." "Smith's sleight of hand he himself called the "invisible hand" . . . What appears to individuals as their own voluntary activity is used, cunningly, by nature to harmonize the whole, so that each person is 'led by an invisible hand to promote an end which was no part of his intention' (*The Wealth of Nations*, 4:2: 485)."

12 A part that functioned to control the functioning of the other parts, which was itself associated with the political notion of "wise government of a complex entity to promote the general welfare" (Canguilhem 1988: 131).

propelled by the life force of capital accumulation. The requirements of this life force structure the relationship of parts within the whole, ordaining the extraction of surplus value from labor by capital, for example, which is facilitated by the division of functions among financial, commercial, and industrial capitalist fractions. Social labor is pumped from the industrial heart of the economy and circulates through the veinous circuitry in its commodity, money and productive forms. As it flows, it nourishes the body and ensures its growth.

As the invisible life force of the capitalist economy, capital accumulation establishes the economy's overarching logic or rationale, its telos of self-maintenance and expanded reproduction. In addition, a regulatory mechanism such as the rate of profit, or competition, or the business cycle, may operate like a thermostat to maintain the economy in a steady state. Ultimately, however, the life "narrative" of the economic organism incorporates not only health and stability but illness and death. Thus, a capitalist economy experiences growth punctuated by crises, and may even be susceptible to breakdowns of an ultimate sort. When it eventually fails and dies, it will be succeeded by another organic totality, a socialism that is presumably better adapted to the conditions that brought about capitalism's dissolution.

Some Marxian theories have attempted to dispel or attenuate the economic determinism and functionalism of this story by externalizing the regulatory function and by theorizing reproduction as a contingent rather than a necessary outcome of capitalist existence. French regulation theory and social structures of accumulation (SSA) theory,[13] for example, have invoked the role of political and ideological – as well as economic – norms, habits and institutions in the process of economic regulation and have attributed to historical "accident" the maintenance of stability in the relation of production to consumption. Despite these attempts to suppress both the teleological and functionalist aspects of "classical" Marxian theory, these frameworks represent the economy and society as an organic structure that operates as a unity among harmoniously functioning parts (see chapter 7). Capitalist history is portrayed as a succession of such structures, each one experiencing maturation and healthy functioning followed by sickness and death. Growth and reproduction are the narrative constants of capitalism's story, revealing the hidden role of accumulation as its life force.[14]

[13] Founding texts within these traditions include, respectively, Aglietta (1979) and Gordon et al. (1982).

[14] This is not to say that all Marxian theorists conceptualize the economy as a coherent and self-reproducing totality but simply that this is a prominent strand of thought within the Marxian tradition (which could be seen as quite internally divided with respect to this type of economic representation.)

In all these narratives there are elements of what might be called cybernetics or systems theory, as well as images of living bodies and machines; indeed, it is difficult to trace concepts like feedback, equilibrium, regulation, and reproduction to a single origin in a particular type of being or science. Though it may be the case as Haraway asserts that the mechanical and cybernetic images became more prevalent in the twentieth century, there was no unilinear movement from organic to mechanical and then to cybernetic conceptions.[15] Thus the concept of the "organism" was not an obvious or natural characterization of the human or animal body, which was developed and then applied to other totalities susceptible to this conceptualization. Rather it was constructed in an interaction of metaphors of economy, machinery, and physiology and indeed only coalesced, according to Foucault, as a hegemonic metaphor of totality, informing both the social and natural sciences, at the end of the eighteenth century and beginning of the age of modernism and of Man.

Metaphor and mastery, organism and intervention

Foucault places in a transitional moment at the end of the eighteenth century the first use of organic structure as a "method of characterization" that

> subordinates characters one to another; . . . links them to functions; . . . arranges them in accordance with an architecture that is internal as well as external, and no less invisible than visible. (1973: 231)

Man's body, constituted as an organism structured by a life force that produces order from within, became at this time the modern *episteme*, setting unspoken rules of discursive practice that invisibly unified and constrained the multifarious and divergent discourses of the physical, life, and social sciences. Modern economics is grounded in Man's body,

[15] While different in detail and language, the structure of the organic and mechanical metaphors is similar, with the entity internally ordered around a hierarchy of functions. Freud speaks in *Civilization and Its Discontents* of the extension of the human organism's powers by the use of tools and machinery: "With every tool man is perfecting his own organs, whether motor or sensory, or is removing the limits to their functioning" (1930: 27); "man has, as it were, become a kind of prosthetic God. When he puts on all his auxiliary organs he is truly magnificent . . . " (pp. 28–9). The permeable boundary between body and machine is one of the things that allows the easy translation between organic and mechanical imagery that is so characteristic of economic discourse today. It is often in the context of Keynesian policy discussions, which are more accommodating to the role of a driver, that the mechanical representation replaces the organic – thus the familiar images of getting the economy rolling again, kick-starting it, etc.

finding the essence of economic development in man's essential nature – his labor (the struggle against nature and death), for example, or his needs and desires (Amariglio 1988: 596–7; Amariglio and Ruccio 1995b). These bodily essences structure a field which is itself the very map of Man, an economy that is organically interconnected, hierarchically organized and engaged in a process of self-regulated reproduction.

Feminist theorists have argued that it is a gendered body "that was the foundation for representing all things, and thus giving things their hidden meaning" (Amariglio 1988: 586) in the modern age. In the modernist regime of gender, human characteristics and other categories are disaggregated upon a binary discursive template in which one term is dominant and the other subordinate and devalued. Though the two terms exist in and through relation to each other, the regime of gender conveys a license to forget the mutuality of dependence. The dominant term thus becomes independent – in other words, its dependence upon its other for its very existence is forgotten – while the subordinate term is unable to exist without its opposite; it is defined negatively, as all that the dominant term is not.

It is not difficult to see in the story of Man and his body the interplay of an infinite set of gendered oppositions – a brief list might include mind/body, reason/passion, man/nature, subject/object, transcendence/immanence. What is interesting, however, is the way in which the regime of gender is a *colonizing* regime, one that is able to capture other dualities and to partially subsume them. Thus as soon as we produce a dualism incorporating two related terms, gender may operate to sustain meanings of wholeness, positivity, definition, dominance, reason, order, and subjectivity (among others) for the first term and incompleteness, negativity, unboundedness, subordination, irrationality, disorder, and objectification for the second.[16]

In this way it becomes possible to understand the bizarre dance of dominance and submission through which Man addresses the economy. When Man is positioned as the first term in their binary relation, he is the master of the economy and of its processes; but when Man (perhaps in the guise of "society") is positioned as the second term, he bows to

[16] The colonizing aspect of the regime of gender has to do with its embeddedness in what Derrida calls the metaphysics of presence in which true identity (or presence) involves exclusion and demotion (of the absent, or what it is not). Feminist theorists, including Irigaray and Kristeva, have long argued that this metaphysics is "implicitly patriarchal; the very structure of binary oppositions is privileged by the male/non-male (i.e. female) distinction" (Grosz 1990a: 101). One could also say, however, that it is racist, heteronormative, and many other things; in other words, it is not necessary to privilege gender in the construction of identity/presence and the consequent devaluation of difference or the "other."

the economy as to his god. Each positioning is informed and constituted by an infinity of binary hierarchalizations.[17]

Man and economy are related by analogy and conflation as well as by hierarchical opposition. Each is a body governed by Reason or a locus of Reason in an irrational domain. Each is an organic unity that maintains itself by subsuming or displacing its exterior, producing integration and wholeness as an effect.

In Man's discursive constitution, dominant (male) human characteristics are represented as universal while subordinate (female) characteristics are externalized or suppressed. They subsist as the Other – woman or nature – to Man, by whose absence or suppression he is defined. Through the operation of the regime of gender, Man becomes a creature who is fundamentally rational and whose fate is mastery and control – of nature, of woman, of all non-Man (Sproul 1993). He is the arrogant knower, whose thoughts replicate and subjugate the "real."

By analogy and by extension, the economy is the locus of Reason in the social totality; it is therefore the dominant social instance. It is the social site of rationality and order, to which the irrational disorder of non-economic life must submit. This hierarchical ordering of the social body, with the economy at/as its head, can be translated into relations of determination. The economy's ability to author its own causation – and to produce its own wholeness and sufficiency as an effect – confers upon it the status of *determinant* with respect not only to itself but to its insufficient other, the external *determined*. Thus the organismic conception contributes to the emergence and prevalence of economic determinism, positing the non-reciprocal relation of economic cause to social effect. As Man is the subject of history, and all the world his object, the economy is the subject of society and enacts its effects upon that passive terrain.

Man and the economy are masters of themselves and of their external domains, and it is through Reason that their internal and external mastery is attained. The analogous operation and dominance of Reason in both beings guarantees the truth of rationalist economic knowledges and techniques. Through Man's logical powers, the orderly operations of the economy can be mirrored, its functioning preempted by his deductions. It is this subjective conflation that gives Man the organic knowledge

> to invent forms of production, to stabilize, prolong, or abridge the validity of economic laws by means of the consciousness he attains of them and by means of the institutions he constructs upon or around them, . . . (Foucault 1973: 369, speaking about the historicity of man)

[17] E.g., man(mind)/economy(body) or economy(god)/man(humanity), ad infinitum.

Given its qualities of wholeness, transcendence, and rationality (for which one might read "perfection") the organic economy is sometimes seen as functioning appropriately without intervention. From certain perspectives, the economy is the word to which the flesh is always and necessarily subsumed. From others, the existence of reason in the economy signals the possibility of successful intervention but also and simultaneously the limited need and scope for intervention. Thus the economy may need its "pump primed" or its life force "re-ignited"; it may need to be "whipped into shape" or "kick-started" to get it "rolling" again. Someone may need to take the helm, pulling on the "levers" that govern the speed and direction of the machine:

> ("Mr Keating emerges from his bunker"): headlines shouted that he was picking up the reins, handling gears and pulling levers again. (Morris 1992: 24)

> Once Labor was elected, the labour movement made a number of assumptions about taking control of the economic levers of power. (Comment by Chris Lloyd, a left-wing union researcher, from an interview by Curran 1991: 27)

Ultimately, however, these interventions are subservient to the logic and functioning of the economy itself.

Finally, there are those for whom the determinist logic of the economy, and its replicability in the rationalist formulations of the mind, make possible the invention of model economic experiments, rationally operating creatures wholly sprung from the mind of Man. These often represent the economic organism transmuted into the noncapitalist form of socialist or libertarian utopias.

In all these conceptions, the economy is both the master of Man and the site of his mastery, whether that mastery be gained through knowledge or through action. This paradox reflects Man's dual existence: as mind and as embodied Reason, he governs and controls; but as mere and mortal body, he looks to the economy, the perfect face of Reason, and submits to it as to his god. This back and forth is the signature of the binary and hierarchical regime of gender. Man cannot escape it, for it is his creator. Instead he plays it out in the discourse and practice of economic intervention.

Bypass surgery: tinkering with the ticker

The organismic economy calls forth a particular discourse of intervention that establishes the masculinist subject position of intervener/controller. Thus the affective discourse of economy is always to some extent a discourse of mastery: the terrain of the economy is laid out by economic theory, with its entryways and pathways clearly marked and its systems interconnected. Spreading the economy before him as his dominion, economic theory constructs Man as a sovereign/ruler. And the familiar terrain of the body is his domain.

It is not hard to see lurking in the vicinity of economic and industrial policy a body engaged in a battle for survival. Couched in the language of the living body or machine, the economy is portrayed as an organism (machine) whose endemic growth dynamic (or mechanical functioning) is in jeopardy. Diagnoses usually focus upon two key areas of economic physiology, obstructions in the circulation system and/or malfunctioning of the heart. The faltering national economy is often compared to healthier bodies elsewhere, all poised to invade and deprive the ailing, or less fit, organism of its life force. Economic and industry policy is formulated to remove the internal, and create immunity to the external, threats to reproduction.

The analogy of the blood's circulation system and the role of the heart in keeping the volume and rate of flow sufficient to ensure reproduction enables a specific set of interventions and manipulations. In recent years, for example, in most industrialized nations the call for wage restraint has been justified in terms of the presumed negative effect of wage increases upon profitability and economic growth. Wages, it is argued, have been the problem, the obstruction in the system of capital circulation that has prevented growth. In the United States wage cuts have been implemented through such tactics as union decertification, two-tier wage structures, and concession bargaining. In Australia, federally legislated policies of wage restraint have been supported by the unions through the Accord.[18]

Visions of an organized and interconnected economic system in which interventions have predictable (and even necessary) effects have facilitated the acceptance of cuts in real wages in Australia. Wage increases have been portrayed as blocking (via their influence on the rate of profit) the generation of a pool of funds available for investment in the expansion and modernization of Australian industry. The backwardness of national industry has been seen as the major constraint upon the international competitiveness of Australian products. By the straightforward logic of organic reproduction, in which specific and focused interventions have a noncontradictory and presumably restorative effect on the whole, wage cuts have been proposed not only to free up investment capital and

[18] The Accord is the tripartite agreement established in the early 1980s between the newly incumbent Federal Labor Government (then under the leadership of Bob Hawke), business interests and established labor unions. In its various incarnations, the Accord has established the guidelines for industry and work practice deregulation and reregulation. It was built upon Hawke's reputed strengths as a conciliator and arbitrator of traditionally opposed interests.

increase competitiveness, but to "overcome the problem of a deficit in the current account of the balance of payments" by "curtailing the demand for imports" and "cutting the costs of exporting and import-substituting industries" (Stilwell 1991: 32).

When a totality is centered, internally connected, hierarchically ordered and governed by laws of motion that can be replicated by reason in the mind of man, the strategist has only to identify the right place to start the treatment (tinkering) and soon the whole will be healthy (working) again. Curtail wages, it is argued, and the flow of investment into the crucial parts of the body economic will take place. At the base of this curative vision is the metaphoric heart of the economy – manufacturing production. It is here that the life blood of the system, capital, is most efficiently created and it is from this site that it is pumped to peripheral sectors and the unproductive extremities.

Given its presumably critical role in economic development and social well-being, it is not surprising that manufacturing investment has long been a concern on the left. In the US in the 1980s, Bluestone and Harrison's influential book *The Deindustrialization of America* (1982) focused attention on disinvestment in the domestic manufacturing sector, identifying foreign investment by multinational corporations and unproductive expenditures on mergers and acquisitions as its principal causes. In Australia, lack of generative investment in manufacturing has variously been attributed to the unwarranted expansionism of the mining sector or the alluring rewards of speculation.[19]

In the context of the prevalent discourse of manufacturing-centrism, it becomes clear that the organicist notion of a hierarchy of functions within the economy – and specifically the essentialist conception that one or several parts are critical while others are peripheral or supportive – has constrained and directed the possibilities of economic intervention. In this as in other centered formulations, the growth dynamic is perceived as emanating from a single economic location.[20] Manufacturing is viewed as the driver of the economy, and all other parts of the economy (including agriculture, services, government, and households) are seen as ultimately

[19] In the 1980s, the problem was seen to lie less in the alternative conduits that drained investment away from Australian manufacturing than in the volume of investment itself which could be derived from the capital–labor relationship. The Accord, with its focus upon wages and industry policy, was established to remedy this.

[20] In many types of economic theory and industry policy discourse, this location is something other than manufacturing (such as tourism, finance or other producer services) but the effects of producing a centered and hierarchically ordered vision are the same. As long as there is a position in theory for a dominant process or instance, analysts will produce a knowledge and politics oriented toward developing and managing that social site to the exclusion of others.

deriving their growth from growth in manufacturing. These other sectors may contribute to the reproduction of capitalist society but they are not the key to its survival – perhaps because they are seen as not generating surplus value, or because they are viewed as low productivity sectors that do not contribute sufficiently to growth, or for some other reason. Growth in these sectors is portrayed as flab, not the hard muscle required for a taut and terrific body economic:

> in order for the shift of employment to services to be developmental and not become a shift to poverty, we (the United States) must maintain mastery and control of manufacturing production. (Cohen and Zysman 1987: 16)

Many types of economic activity are thus relegated to secondary status as targets for resources and attention.[21]

Indeed the organicist conception contributes to a very familiar hierarchy of policy priorities. While some types of economic activity are seen as essential to social survival, and as therefore necessitous of intervention, others are viewed as frosting on the social cake. Though it may be widely recognized and lamented that child-care and its low wage providers are in difficult economic straits, policymakers will remind us that unless we take care of manufacturing we are *all* up the creek.[22]

Buttressed by the conception of the organism as a self-maintaining self-rectifying body, strategists may argue that restoring growth in key or lead sectors will set the entire economy upon a path of growth or recovery. In this view, the principle of efficiency dictates that interventions be targeted at the critical locations. When economic conditions are dire, intervening to improve child-care centers is like offering a bandaid to a patient with a heart attack.

The interconnectedness of the parts, and the accessible logic of their interconnection, enables intervention at some distance from the problem (symptom). It thus becomes perfectly reasonable to argue that if we want decent child-care centers we must start with productivity increases or wage cuts in manufacturing. It is also acceptable to ignore or to postpone

[21] One of the few interventionist strategies to challenge the productionism and manufacturing-centrism of much industry policy was the London Industrial Strategy. Among the political economists and economic geographers who provided the background economic analyses for this broad-based strategy there appeared to be a genuine willingness to question the role of manufacturing in the economy, the reliability of profitability as an indicator of performance, and the marginalization of unpaid labor and non-market activities in economic discourse (Massey 1988). Industry strategies were formulated for cultural industries, child-care and the retailing sector in London (Greater London Council 1985).

[22] In the more mean-spirited version, it is argued that child-care helps those with children while manufacturing helps us all.

dealing with problems in most parts of the economy since presumably these will be rectified by the healthy functioning of the heart.

The truth of all these representations is guaranteed by a rationalist conviction that the reductive logic of economists reflects the orderly and parsimonious logic of the economy itself. These logics dictate that economic interventions will have predictable and noncontradictory outcomes and they define the relation of policymakers to the economy as that of Man to machine. Thus you may quite easily arrive at the bizarre conclusion that general economic well-being will be enhanced by wage cuts; and by associating this vision with an invincible and deific figure, you may sell this program to an entire nation of wage earners and economic believers.

Matters of life or death

In economic policy discourse, whatever the diagnosis, there is seldom a question that we are dealing with a unitary system, whose future must necessarily involve reproduction that can only be achieved through growth. To return to the anorexia analogy, the economy is an individual whose survival instinct has been waywardly misplaced, and who must now be forced via gentle or rough persuasion to eat and grow. It is not a collectivity of bodies, which in their diversity are variously getting fat, giving birth, dieting, dying, transforming, and coupling as calories pump into and out them in a decentered, almost directionless way. Rather the economy is an organized and purposeful whole governed by laws of survival that cannot be countervened.

The lawful self-regulation of the economic organism dictates that interventions must ultimately serve or operate within the organism's telos of organized growth. Policy then is affected not only by the essentialism of the organic metaphor, which ascribes generative power and causality to certain aspects of the totality and withholds it from others, but also by the functionalism of this conception.[23] The economy is reduced to a set

[23] This functionalism could be seen as another form of essentialism, in that the economy (or society) itself is the "founding totality of its partial processes" (Laclau and Mouffe 1985: 95).

The charge of functionalism has been made against Marxian economics by countless anti-Marxists as well as by some neo-Marxists (e.g., Elster 1982 and Barnes 1992). Elster and Barnes trace functionalism within Marxism to the inappropriate adoption of biological metaphors of organism and reproduction within a social science that values the reflexivity and individualism of human behavior. But this judgment of inappropriate theoretical choices rests upon the assumption that there is an arbiter of appropriateness (the rules of correspondence or coherence notions of truth) or that objectively "better" metaphors could be found. Such an assumption cannot grasp Foucault's idea of an episteme, which sidesteps questions of appropriateness in search of the rules and conditions of possibility of an historically grounded knowledge.

of functional relations that are coordinated by the rules and requirements of capitalist reproduction. Thus no matter whether an intervention is well or ill conceived and managed, its effects are necessarily to perpetuate "capitalism" and capitalist class relations. This invisible prescription circumscribes and constrains even the most left-wing economic proposals and analyses.

Stilwell (1991) argues, for example, that the expected effects of wages restraint in Australia – deflation, reduction of the balance of payments deficit and growth – were easily subverted by the inflationary effects of monopoly pricing, the increased demand for imports from those on non-wage incomes, and the flow of newly created investment funds into paper entrepreneurialism or property speculation rather than production. Stilwell's economy may not have gotten the infusion that the social democrats intended; the actions of "individuals" (functioning according to the logic of individual self-maintenance rather than in their alternative role as parts of the larger economic and social organism) may have betrayed the common interests represented by the body of which they are a part. Yet this did not ultimately threaten the capitalist organism. Lack of effective management resulted in reproduction locally of the ugly face of capitalism – workers with wages cut and no revitalized national economy to show for it. But the organism remains intact because the organicist discourse allows for no other proximate outcome.

Organic functionalism subsumes the future to the contours of the present. But it also precludes envisioning diversity and multiplicity in the consequences of economic intervention. Society as organism is a set of conformable interests in which all benefit from the healthy functioning of the whole:

> Functionalism has been developed on a foundation of organismic metaphors, in which diverse physiological parts or subsystems are coordinated into a harmonious, hierarchical whole. Conflict is subordinated to a teleology of common interests. (Haraway 1991: 24)

Certainly, in Australia, the interests of business and the organized labor movement have been represented by political and union leaders as effectively harmonious:

> *Australia* needs a sustainable high growth strategy that avoids or minimizes the effects of the boom-bust cycles of the past. Metal workers and *all Australians* simply cannot afford a vision of nation building which leads to low growth and another one or two boom-bust cycles during the 1990s decade. (MEWU 1992: 24, emphasis mine)

In the face of this kind of assertion, which is buttressed by a notion of common "national" interests, it is difficult to maintain a sense that any "growth strategy" – indeed, any intervention in a complex totality – will have uneven and contradictory effects.

That the strategic unionism advocated by leftists has so easily been led into strategic functionalism, that is, into advocating policies that help materialize the reproduction of capitalist practices, has long been a matter of concern to those whose economics focuses less upon reproduction and more upon the potential for economic dysfunction (MacWilliam 1989). Bryan (1992) argues, for example, that the Australian left had no business supporting any form of wage restraint, as this only served to shore up the accumulation process and avert, once again, the threat of imminent crisis.

The life/death opposition that lies at the nub of the organic metaphor presents the opportunities for political intervention in the form of a simple duality. If I don't wish to pursue industrial strategies for patching up or resuscitating capitalism, I can upend the analysis and concentrate upon exacerbating the pre-conditions of death. Though most leftists now abjure the millennial goal of promoting "the revolution" by promoting organic dysfunction, organic functionalism has locked them into the alternative goal of promoting capitalist health. In order to create employment and rebuild communities, they must participate in strategies and programs to foster capitalist development, capitalist reindustrialization, and capitalist growth (see chapter 7). Many on the left would like to see an alternative to capitalism, but they face a unitary economy that allows for no such proximate possibility. Their options are to promote the healthy functioning of capitalist economies or to see working people and others marginalized and impoverished. This is not a particularly inspiriting choice, yet its grounding in humanism and organicism is seldom questioned or even brought to light.[24]

Beyond life and death

Donna Haraway argues that if the future is given by the possibility of a past, then an "open future" must rest upon a "new past" (1991: 41–2). This could involve, I would argue, a new conception of totality, one that abandons the organism as we know it. Haraway gives some

[24] Callari (1991) argues that the economistic (and organic) theoretical framework of classical Marxism effectively "economized the political" by focusing political discussion upon the economic conditions under which capitalism would fail, rather than the moral and legal, that is, political, processes which determined the future of capitalist practices. By defining political subjects in terms of their economic interests and positions predetermined by the "closed economic mechanism that constituted capitalism" (p. 203) socialists have been strait-jacketed into the logic of this mechanism, perpetually waiting for the "revolutionary moment."

encouragement that such a discontinuity is possible:

> One is not born a woman, Simone de Beauvoir correctly insisted. It took the political-epistemological terrain of postmodernism to be able to insist on a co-text to de Beauvoir's: one is not born an organism. Organisms are made; they are constructs of a world-changing kind. (1991: 208)

In a similar vein, Foucault prepares the way for a rethinking of totality in non-organic and non-anthropomorphic terms. Having shown how the vitalism of organic structure could not have been thought within the discourse of the sixteenth century and thus how Man's body could not have existed as the "ground for discourse" before the nineteenth century (Amariglio 1988: 589), he speculates in the conclusion of *The Order of Things* upon the end of the modern episteme and the fundamental arrangements of knowledge that made it possible for the figure of Man to appear:

> As the archaeology of our thought easily shows, man is an invention of recent date. And one perhaps nearing its end.
>
> If those arrangements were to disappear as they appeared, if some event of which we can at the moment do no more than sense the possibility – without knowing either what its form will be or what it promises – were to cause them to crumble, as the ground of Classical thought did, at the end of the eighteenth century, then one can certainly wager that man would be erased, like a face drawn in sand at the edge of the sea. (1973: 387)

In a search for a new social and economic totality, born of the old but perhaps not its semblance, I sometimes turn to discourses of economic change.[25] Certainly, I tell myself at these moments, it is in the

[25] The question of course arises whether we want to dispense with the concept of totality entirely. This is certainly an option, but one which leaves the concept untouched, whereas the alternative option of reworking the concept of totality will always be to some extent compromised by the organicist meanings of the term (see Cullenberg 1994b). Each strategy has its strengths and its pitfalls. In this paper, I have chosen to rework rather than abandon "totality" as a concept, taking inspiration from feminist projects of retheorizing the body. Feminist rethinkings of the body (its boundaries, its hierarchical ordering, its psychological and social topography, etc.) have not meant purging the body from discourses of the subject and society; on the contrary, they have been partially responsible for reinstating the body as a prominent focus of such discourses, one with important theoretical implications and social effects.

Laclau and Mouffe are engaged in an interesting project of retheorizing the social totality, though one different from the project I am pursuing here: "Our vision is to a large extent holistic, since it presupposes that any identity is differential . . . and that the systems of differences are articulated in totalities which are 'historical blocs' or 'hegemonic formations.' But unlike classical sociological holism . . . we do not feel these configurations or social totalities to be self-regulating totalities but precarious articulations that are always threatened by a 'constitutive outside'" (Laclau 1990: 221–2).

discourse of economic restructuring, produced over the last twenty years by Marxist political economists in a variety of social science fields, that I have had the most experience of (re)constructing the organic economy. Perhaps it is also in this context that I might have the greatest chance of perceiving an emergent totality,[26] one that is no longer constrained by essentialism and reproductionism, or inflected with the arrogance of interventionist humanism. Perhaps I might find the ground from which to move beyond the outmoded but still unreplaced "progressive" options of socialist "revolution" or capitalism with a human face.

The ladder of evolution

Genealogies of capitalism, metaphors of organic development

The discontinuity which, in Foucault's archaeological terms, marked the beginning of the modern age brought the rise of History as the organizing principle of knowledge. Along with History came an interest in the internal organic relations between elements of a totality, the life and death of organic structures, and the linear sequencing, or succession, of analogous structures (1973: 218–19).

Certainly in the discourse of economic change there has been no shortage of coherent structures succeeding each other in orderly progression. In recent years, for example, one of the distinctive features of Australian left-wing industrial policy has been the promotion of a new "model of industrial development." This model is none other than post-Fordism, an industrial "paradigm" that focuses upon the developmental role of small and medium-sized firms and the reorientation of business and work cultures around flexibility, computerized technology, networking, and strategic alliances both within sectors and between producers and consumers (Mathews 1990). The aim of industry interventions is to create the conditions under which a fully fledged post-Fordist economy might be born, unimpeded by obstructionist union regulations or demarcations, business attitudes, or statutory barriers. Underlying the vision of the new industrial model are the familiar metaphor of the economic organism

[26] The literature on internationalization is a good example of a discourse that constantly undermines the notion of organic boundary. One of the difficulties still faced in this literature is the problem of how to replace the conception of a "national economy" (a bounded organism) with any meaningful unit. While some political economists have substituted international capitalism as the mega-organism (Bina and Yaghmaian 1991; McMichael and Myhre 1991), others have abandoned the search for a self-reproducing, holistic totality in favor of an overdetermined totality of processes (capitalist and noncapitalist) that occur over space (global and non-global) (Ruccio, Resnick and Wolff 1991; McIntyre 1991).

and an associated conception of capitalist development as a succession of organic structures, or "models of development" (this term is taken from Lipietz 1992), each structurally similar to but qualitatively different from the last.[27]

In his collection of "popular scientific" essays on origins and evolution, Stephen Jay Gould (1991) tells the wonderful story (entitled "Life's Little Joke") of competing depictions of the evolutionary development of the modern-day horse. Until recently, the case of the horse has served as the common illustration of species evolution up a ladder of continuous development from primitive to modern. Each lock step of the ladder is marked by increasing size and height, decreasing number of toes and an increase in the complexity of the grinding teeth. This standard iconography of evolution has, according to Gould, "initiated an error that captures pictorially the most common of all misconceptions about the shape and pattern of evolutionary change" (p. 171). The metaphor (and illustrative device) of a ladder portrays evolutionary development as an unbroken continuity. It encapsulates the view that horses developed through a series of sequential stages of development, each adapted to the changing environment at hand. In similar fashion, the current penchant for representing the history of twentieth-century capitalist development in terms of a series of progressive steps from pre-Fordism to Fordism to post-Fordism places economic organisms on a ladder of sequential adaptation (see figure 5.1).[28]

Gould's reading of the fossil evidence, and that now commonly accepted, has caused a radical rethinking of the ladder metaphor and the adaptive functionalism it embodies. He argues that the metaphor of a bush might better suit the evolutionary drama that is partially revealed by the fossil record:

> Evolutionary genealogies are copiously branching bushes – and the history of horses is more lush and labyrinthine than most. To be sure,

[27] In *Working Nation: The White Paper on Employment and Growth* recently issued by the Labor government in Australia, not only is a post-Fordist model of development represented as the optimal way forward but the body of the economic region has undergone a marked transformation. No longer starved and anorectic, in need of force-feeding with infusions of outside investment, the regional economic body is now pregnant with possibility: "[The post-Fordist model] portrays the region as already full of economic potential that needs only to be liberated by intervention" (which will mainly take the form of) instilling a new business culture within local areas . . . Now regions are the homes of 'stakeholders' who have it in their power to make their regions into 'pockets of excellence,' 'entrepreneurial hotspots,' in short, industrial growth centres" (Gibson-Graham 1994a: 149).

[28] Alternatively, capitalist development has been theorized as a succession of social structures of accumulation, regimes of accumulation and modes of regulation, or as the supersession of organized by disorganized, or competitive by monopoly, capitalism.

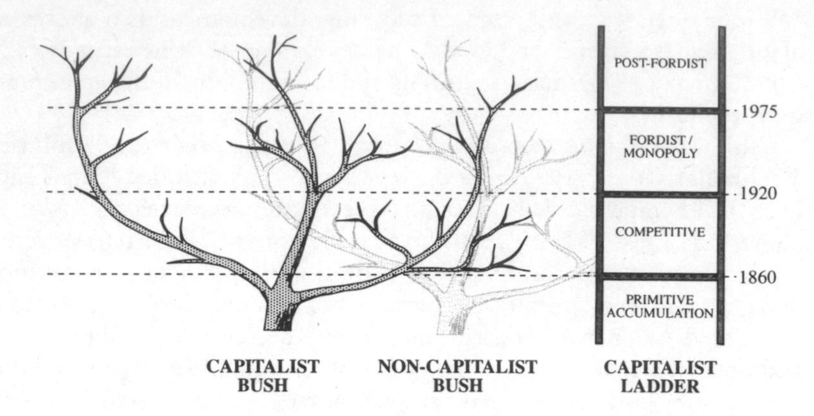

Figure 5.1: Metaphors of economic evolution

> Hyracotherium is the base of the trunk (as now known), and Equus is the surviving twig. We can, therefore, draw a pathway of connection from a common beginning to a lone result. But the lineage of modern horses is a twisted and tortuous excursion from one branch to another, . . . Most important, the path proceeds not by continuous transformations but by lateral stepping . . . (Gould 1992: 175)[29]

Within economic restructuring discourse some empirical studies likewise question the hegemony of the ladder of economic development. Storper (1991), for example, has produced an interesting discussion of four different models of technically dynamic industrial development that have coexisted during the twentieth century within different cultural contexts. Only one of these models (found, not surprisingly, in the United States) is consistent with what we have come to call Fordism.[30] Piore and Sabel (1984) have highlighted the viability of forms of flexible specialization within capitalist industry in northern Italy throughout the so-called Fordist era. The work of economic sociologists and anthro-

[29] The imposition of the model of a ladder upon what, in Gould's reading, is "the reality of bushes" places at the forefront of evolutionary progress only unsuccessful lineages on the very brink of extinction "for we can linearize a bush only if it maintains but one surviving twig that we can falsely place at the summit of a ladder" (p. 181). The familiar iconography of evolution shows, then, rather than a ladder of progressive adaption and evolution, a pathway to extinction. Life's little joke is that humankind is often portrayed at the pinnacle of a similarly structured hierarchy of living things, highlighting, for Gould, the imminence of our species extinction rather than our evolutionary superiority.

[30] Criticism of the generalizability of the (US based) Fordist mass production industrial paradigm has come from many quarters. Hudson and Sadler (1986), for example, have questioned its relevance in the UK.

pologists suggests a vision of a diversity of industrial structures, firm types and models of development interacting in different combinations. The selection of particular models as "universal" or "dominant" in the accepted narratives of capitalist development reflects, I would argue, the power of metaphors of organicism and ladders of evolutionary change.

In economic development theory as in biology there has been a tendency to run "a steamroller over a labyrinthine pathway that hops from branch to branch through a phylogenetic bush" (Gould 1992: 180) of economic forms (see figure 5.1). In the process the many capitalist and noncapitalist forms that have co-existed with the "dominant" form have been obliterated from view. This discursive marginalization functions powerfully to constrain the visions and politics of the future, prompting, for example, industry interventions designed to facilitate the step into post-Fordism (seen as currently the most adaptive, advanced, and efficient form of capitalism) and thereby making it less likely that non-post-Fordist and noncapitalist forms will continue to exist (see chapter 7).

As Gould's story shows, the representation of history as a sequential ladder has the effect of reducing eco(nomic)-diversity. By denying the existence of other branches and pathways, the image of development as a ladder of evolution promotes the monolithic capitalism it purports to represent. In its most egregious and easily recognizable manifestation, the development ladder ranges the countries of the world along a unilinear hierarchy of progress, calling forth attempts to eradicate "traditional" economic forms and replace them with capitalist industrialization.[31]

Modern Darwinian evolutionary theory constructs a vision of the "naturalness" of domination. During the early nineteenth century, the representation of the body or population (animal, vegetable, or human) as an organism which is somehow internally motivated by a fight for survival became inextricably linked to concepts of natural dominance (Haraway 1991: 42). In economic terms, dominance came to be understood as the dominance of capitalism and capitalist class processes over all other forms of economy and exploitation. Economic evolution has become a story of the progressive emergence of ever more efficient, more competitive, and therefore dominant forms of capitalist enterprise, technology, and economic organization.

[31] Of course it is useful to remember that the ladder metaphor plays not only a central role in the economic development literature, but also in treatises about socialist transition. Socialism has often been seen as the lock step above capitalism in the development ladder, a vision that has now lost most of its potency, even on the left.

In rethinking the economic totality, perhaps we might begin by abandoning the hegemony of dominance, as some feminist theorists have begun to do. Perhaps we might also abandon the narrative of History as a succession of hegemonic structures, each of which has won a war of survival and adaptation.[32] Finally and most importantly, we might abandon the organic body economic and seek a "new conception of the organism as an intereffective totality of determinations," as Richard Lewontin puts it (quoted in Amariglio, Resnick and Wolff 1988: 499), or something analogous on the social level.

In an "intereffective social totality" each economic process might be understood as overdetermined by all non-economic processes, and as participating in their overdetermination (Resnick and Wolff 1987).[33] Privileged economic sites and processes would thereby lose their status as causes that are not simultaneously effects. Lacking its unifying rationale or essential life force, the economy would be deprived of its integrity and its commitment to reproduction. As the desiccated shell of the organism fell away, we might glimpse a region of infinite plurality and ceaseless change, in which economic processes scatter and proliferate, unhampered by a ladder of development or a telos of organized growth.

Here again Gould's story may contribute to a reconceptualization:

> Who ever heard of the evolutionary trend of rodents or of bats or of antelopes? Yet these are the greatest success stories in the history of mammals. Our proudest cases do not become our classic illustrations because we can draw no ladder of progress through a vigorous bush with hundreds of surviving twigs. (1991: 180)

My analogous question is "Who ever heard of the development in the contemporary western world of noncapitalist class processes[34] like feudalism or slavery as prevalent forms of exploitation, or of independent commodity production as a locus of "self-appropriation"? Yet these are the greatest survival stories in the history of class. Our focus on the development of the different forms of capitalist enterprise (and by implication

[32] As poststructuralist and some forms of post-Marxist theory urge us to do (e.g., Laclau and Mouffe 1985). Consider, for example, the plea of Soja in his essay on "History: Geography: Modernity" for a liberation of the geographical or spatial imagination from "an overdeveloped historical conceptualization of social life and social theory" (1989: 15).

[33] The concept of overdetermination (see chapter 2) involves the mutual constitution and intereffectivity of all social and natural processes. This concept allows for a decentered vision of social sites and a nondeterminist reading of historical eventuation.

[34] By class process I mean the process of producing, appropriating and distributing surplus labor which involves an exploitative moment (in which surplus labor is appropriated from its direct producer) and a distributive moment in which it is distributed to various social uses and destinations (see chapter 3).

of capitalist exploitation) has made it difficult to conceptualize the persistence and establishment of many noncapitalist forms of exploitation in households, shops, small factories, farms and communes (represented in figure 5.1 as a shadowy bush). Our metaphor of the organism, in its functionalism and holism, has contributed to the portrayal of all noncapitalist class processes as subordinate to and reproductive of "capitalism." It has fostered an understanding of capitalism as a unitary figure coextensive with the geographical space of the nation state (if not the world)[35] rather than as a disaggregated and diverse set of practices unevenly distributed across a varied economic landscape. On the metaphorical ladder of evolutionary development, noncapitalist forms of exploitation have been denigrated as primitive remnants of a dominance long past, perhaps still existing in Third World countries but not consequential in the social formations of the so-called developed world. Ignored by socialists focused and fixated on capitalist dominance, these noncapitalist forms have been neglected as sites of political activity and class transformation or dismissed as the revolutionary ground of populists and romantics.

No organism, no guarantees

By centering the organic economy on capitalist class processes and on ostensibly dominant economic forms, economic policy discourse curtails and truncates the possible avenues of economic intervention, to the cost of all those interested in the political goal of class transformation (Ruccio 1992). The ladder of development that places post Fordism (or some other successful form of capitalism) at the pinnacle of contemporary economic adaptation precludes the possibility that noncapitalist adaptation may be simultaneously taking place and, at the same time, precludes the possibility of successful socialist projects and interventions.

In the face of this restrictive vision and the set of possibilities it allows, some feminist theorists have abandoned the conception of the economy as a unified and singular capitalist entity, emphasizing the role of the household as a major site of noncapitalist production in so-called advanced capitalist social formations (see, for example, Folbre 1993, Waring 1988). Eschewing the formulations of what is sometimes known as dual systems theory, in which patriarchy and capitalism are viewed as two forms of exploitation situated respectively in the household and industrial workplace, certain feminist theorists have identified a variety of forms of household class relations (Fraad et al. 1994; Cameron 1995).

[35] This conception is certainly the distinctive and most powerful legacy of classical economics.

They represent the household as a site of difference and change in terms of both the types of production that take place there (including use values for domestic consumption, like clean rooms and cooked meals) and the ways in which surplus labor is produced and appropriated by household members.[36]

This feminist attempt to retheorize and displace "the economy" has powerful and potentially far-reaching implications. It effectively decenters the discourse of economy from the capitalist sector without at the same time establishing an alternative center for economic theory. At the same time, its emphasis on the diversity of *household* forms of economy and exploitation opens the possibility of theorizing class diversity in the *non-household* sector. Once that possibility exists, we may begin to produce a knowledge of diverse exploitations in "advanced capitalist" social formations. Such a knowledge is one of the conditions of a politics of class diversity, and the absence of such a knowledge is one of the conditions that renders such a politics unthinkable and obscure.

The hegemony of the organism and the ladder within certain types of Marxian (and much non-Marxian) economic theory has prevented a complex, decentered knowledge of an overdetermined economic and

[36] Some have argued that, in certain households, the feudal domestic relation (see chapters 3 and 9) in which a woman produces use values that are appropriated by a male partner is being politically renegotiated under the influence of feminism and the growing acceptance of gender equality (e.g. Fraad et al. 1994). In these households communal processes of surplus labor production and appropriation are being invented and explored. It is interesting to note that many industry interventions are actively undermining the viability of noncapitalist class processes – both within and outside the household – rather than supporting them. In the push to establish post-Fordism in Australia, for example, the centralized wage fixing system that has prevailed throughout most of the twentieth century is being dismantled. The move to enterprise bargaining threatens to destroy the established tradition of flow-ons whereby the gains of the organized labor movement have been generalized across the economy as a whole. Negotiation of a communal class process in households rests, in part, upon the growing economic independence and equality of women *vis-à-vis* their male partners. In Australia, at least, any trend toward gender wage equity, or comparable pay, has come through industry union representation. Under this system many women workers have earned industry standard award wages and regulated working conditions that have helped to secure their economic rights in household negotiations. The adoption of enterprise bargaining has little to offer most women as, in the deregulated but still segregated labor market, it is they who are often employed in smaller, more risky companies in which their bargaining positions may be weak. Indeed the first Annual Report on Enterprise Bargaining (1994) reveals that women covered by certified agreements and nonunion deals are less likely than men to receive wage increases (Martin 1995: 4). The class effects of such an industry policy may well be to resuscitate the capitalist class process of surplus value production and appropriation within capitalist enterprises, large and small, while at the same time undermining one of the conditions of existence of communality in households.

social totality from emerging. These metaphors have generated a simple and restrictive vision of "the economy," one that – in ironic counterpoint to the assessed failure of most economic policies and programs – is associated with a discourse of masterful intervention and mechanical eventuation. To the extent that this vision has currency, economic discourse and the economic policy it gives rise to is a drama in which Man aspires to the state of transcendent Reason and mastery. Unfortunately, arrogance and failure are the shadows that play upon the stage.

To envision the economy as an overdetermined social location, no more susceptible to logical or active mastery than is the world in its contradictory fullness, proliferative rather than reductive of forms, profoundly unstable yet immoveable from the fulcrum of economic intervention, is to forego the ecstasy of rationalism and the arrogant security of determinate effects. Yet it is also to give up the organic totality and its linear path of evolution and to see beyond reproducing capitalism with a human face.

Though it is not therefore malleable to our manipulations, our totality is what we discursively make it. Perhaps we can make it a site for the envisioning and enactment of new class futures.

6

Querying Globalization

It was an article on rape by Sharon Marcus that first "drove home" to me the force of globalization. The force of it as a discourse, that is, as a language of domination, a tightly scripted narrative of differential power. What I mean by "globalization" is that set of processes by which the world is rapidly being integrated into one economic space via increased international trade, the internationalization of production and financial markets, the internationalization of a commodity culture promoted by an increasingly networked global telecommunications system. A forceful visual image of this present and future domain is the photograph of Spaceship Earth which is increasingly used to advertise the operating compass of global banks or businesses, promoting the message that "we" all live in one economic world.[1]

Heralded as a "reality" by both the right and the left, globalization is greeted on the one hand with celebration and admiration, on the other with foreboding and dismay. This chapter focuses initially upon left discussions in which globalization is represented as the penetration (or imminent penetration) of capitalism into all processes of production, circulation and consumption, not only of commodities but also of meaning:[2]

The distinctive mark of late capitalism, underpinning its multinational

[1] The photographs of "Spaceship Earth" taken during the Apollo space missions from 1968–72 have played a powerful role in envisioning one world. They ushered in the eras of both "environmental crisis" and "global capitalism" and have been deployed to popularize environmental concerns for "life on a small planet," on the one hand, and to celebrate the technological and economic triumphs of global corporate integration on the other. Geographers Cosgrove (1994) and Roberts (1994) provide interesting readings of the uses of these images in discourses of globalization.

> global reach, is its ability to fuse and obliterate the boundaries of production and consumption through the *pervasive domination* of all levels of consumer culture and everyday life. (Deshpande and Kurtz 1994: 35 emphasis mine)

> Multinational capital formation . . . no longer makes its claims through direct colonial subjugation of the subject, but rather by the hyperextension of interpellative discourses and representations generated with and from a specifically new form of capital *domination*. Thus, it is important to recognize that *domination* occurs intensively at the levels of discourse, representation, and subjectivity. (Smith 1988: 138 emphasis mine)

> the prodigious new expansion of multinational capital ends up *penetrating* and colonizing those very precapitalist enclaves (Nature and the Unconscious) which offered extraterritorial and Archimedean footholds for critical effectivity. (Jameson 1991: 49 emphasis mine)

Fueled by the "fall of socialism" in 1989, references are rampant to the inevitability of capitalist penetration and the naturalness of capitalist domination. The dynamic image of penetration and domination is linked to a vision of the world as already or about to be wholly capitalist – that is, a world "rightfully owned" by capitalism.

It was in the light of this pervasive discourse of capitalist penetration and of assertions that the Kingdom of Capitalism is *here* and *now* that Marcus's argument about rape spoke to me so directly. Marcus challenges the inevitability of the claim that rape is one of the "real, clear facts of women's lives" (1992: 385). She draws attention to a "language of rape" which assumes that "rape has always already occurred and women are always either already raped or already rapable" (1992: 386). Reading her analysis of the positioning of women in the language of rape, I found myself replacing her object of study with my own – rape became *globalization*, men became *capitalism* or its agent the *multinational corporation (MNC)*, and women became *capitalism's "other"* – including economies or regions that are not wholly capitalist, non-commodified exchanges or commodities that are not produced by

2 Where globalization is seen in terms of penetration, the parallels with rape are obvious. I wish to argue that celebratory admiration of globalization is not limited to the right and that, despite the criticisms of globalization that emanate from the left, there is an undercurrent of leftist desire for "penetration." Consistent with the often expressed view that the one thing worse than being exploited by capital is not being exploited at all, there is a sense that not to be penetrated by capitalism is worse than coming into its colonizing embrace. The ambivalent desire for capitalist penetration is bound up with the lack of an economic imaginary that can conceive of economic development which is not *capitalist* development (with its inherent globalization tendency), just as conceptions of sexuality that are not dominated by a phallocentric heterosexism (in which the act of penetration, whether called rape or intercourse, defines sexual difference) are difficult to muster.

capitalist production, and cultural practices that are conducted outside the economy.

When Marcus expressed her concern that feminist activism to make rape visible in communities had also produced a politics of fear and subjection, I thought of the industry activism I had participated in that had generated a knowledge of globalization and capital flight, and the way in which a politics of fear and subjection had emerged from it. When she made reference to the ways in which women limit their activities – for example, avoiding or thinking twice about being out in public spaces alone or in the evening, for fear of being accused of "asking for" rape – my mind turned to the way in which workers have limited their demands for higher wages or improved working conditions, given the knowledge about capital mobility and the operations of the MNC, for fear that they might be "asking for" capital abandonment.[3]

But how, Marcus prompted me to ask, was this message of fear conveyed by analyses of globalization? How had globalization become normalized so as to preclude strategies of real opposition, as in Marcus's view the normalization of rape has hindered feminist strategies for real and active resistance within the rape event itself? Within the discourse of globalization fear of capital flight and subjection to heightened exploitation are positioned as legitimate responses to the so-called "realities" of globalization[4] – just as fear of violation, and passivity or paralysis in the face of violence become legitimate responses to the "realities" of rape. As Marcus notes, "(w)e are taught the following fallacy – that we can best avoid getting hurt by letting someone hurt us" (p. 395). She argues that unquestioned acceptance of rape as a fact of life limits the political efficacy of feminist actions that could be taken to prevent it. Accepting men's capacity to rape as a self-explanatory "fact" encourages activism in the areas of litigation and reparation "after the fact," rather than in the form of direct challenges to the victim role prescribed by the "rape script."

[3] It is not hard to see how the knowledge of inferior wages and working conditions in a distant industrial setting, conjoined with the pervasive belief in the driving force of international competition, global economic integration and the power of MNCs, could have a disciplining and dampening effect upon local workplace struggles.

[4] And here, it might seem, the analogy between rape and global economic processes could be seen to founder. For women, in this argument, the fear is of bodily penetration while, for workers in the case I alluded to, the fear is of capital withdrawal. I would argue that in the discourse of globalization, capital flight is simply the flip side of a process of increasing capitalist penetration and integration on a global scale. Withdrawal of capital from one productive site at one geographical location does not leave that space empty of capitalism – it does not, for example, rule out capitalist activity in that same location at another economic site, such as the service sector – and thus does not indicate any less penetration.

Parallels can be drawn here with the 1970s industrial activism that accompanied the widespread acceptance of capitalist globalization as a "fact." Some unions within national labor movements were tempted into aggressive bids for their piece of the action – a role in the World Car plan, for example. Others accepted globalization by pressing for state assistance to regions hardest hit by capital flight. Rather hollow-sounding calls for the "international workers of the world" to unite were revived; but while the polemic was inspirational for a time (in a nineteenth-century kind of way) efforts to realize such a movement were largely thwarted by the discursive positioning of capital's globalization as a more powerful force than labor's internationalism.[5]

In her suspicion of claims about "reality" and the unmediated nature of women's "experience," indeed, in her refusal to recognize "rape as the real fact of our lives" (1992: 388), Marcus is led to explore the construction of rape as a linguistic artifact and

> to ask how the violence of rape is enabled by narratives, complexes and institutions which derive their strength not from outright, immutable, unbeatable force but rather from their power to structure our lives as . . . cultural scripts. (1992: 388–9)

A language shapes the rape script – "the verbal and physical interactions of a woman and her would-be assailant" – permitting the would-be rapist to constitute feelings of power and causing the woman to experience corresponding feelings of terror and paralysis. But the language also "enables people to experience themselves as speaking, acting, embodied subjects" (p. 390), offering the potential for new words, different feelings and unexpected actions to emerge. By invoking the metaphor of a language, something that has rules of grammar but rules that can be used to say many different things, Marcus suggests that we might begin to imagine an alternative script of rape (1992: 387).

I could see in Marcus's project the contours of an argument very similar to the one I was constructing to challenge the dominant discourse of capitalist globalization. And what began as a simple game of seeing how far the comparisons and analogies between rape and globalization might take me ultimately developed into a rather complex form of play

[5] Herod documents this positioning in the industrial restructuring literature, commenting that "there is little sense that workers are themselves capable of proactively shaping global economic landscapes through their direct intervention in the geography of capitalism. They are portrayed as the bearers of global economic restructuring, not as active participants in the process" (1995: 346). The close relationships between many left political economic theorists of globalization and labor unions suggests that the hegemonic representations of globalization were not unimportant in shaping the rather constrained politics of opposition that emerged in response.

– the project of reading the globalization literature through the lens of poststructuralist feminist and queer writings about bodies, sexuality and gender. In pursuing this project I hoped to gain some purchase on globalization discourse, to become more active in the face of it, less prey to its ability to map a social terrain in which I and others are relatively powerless and inconsequential.

Rape script/globalization script

Marcus conveys many different meanings through her idea of a rape script. She does not use the metaphor literally to invoke a set of preconstituted parts – victim and rapist – in which actors play out fixed roles. Instead she employs the more fluid image of a script as a narrative – "a series of steps and signals" (p. 390) – whose course and ending is not set. Her notion of a script involves the continual making and remaking of social roles by soliciting responses and responding to cues, and in this sense she highlights the self-contradictory nature of any script and the ways in which it can be challenged from within (pp. 391, 402). When she narrates the common rape script she draws our attention to the choices that are made, through which unset responses become set and woven into a standardized story.

The standardized rape script goes like this: men are naturally stronger than women, they are biologically endowed with the strength to commit rape. In the gendered grammar of violence, men are the subjects of violence and aggression. Their bodies are hard, full and projectile.[6] Women are naturally weaker than men. They can employ empathy, acquiescence, or persuasiveness to avert (or minimize the violence of) rape, but they cannot physically stop it. In the gendered grammar of violence, women are the subjects of fear. Their bodies are soft, empty, vulnerable, open.[7]

There are many obvious points of connection between the language of rape and the language of capitalist globalization. Feminist theorists have drawn attention to the prevalence of shared key terms, for instance – "penetration," "invasion," "virgin" territory – in commenting on the phallocentrism of most "developmentalist" theory (Fee 1986). But beyond the by now familiar gender coding of the metaphors of economic development, there are interesting resonances in the ways a scripted

6 Marcus paraphrasing Brownmiller: "The instrumental theory of rape . . . argues that men rape because their penises possess the objective capacity to be weapons, tools, and instruments of torture" (1992: 395). See chapter 4 for a discussion of the spatiality of women's bodies.

7 "The rape script describes female bodies as vulnerable, violable, penetrable, and wounded" (Marcus 1992: 398).

narrative of power operates in both the discursive and social fields of gendered and economic violence:

> The most appropriate metaphor for modernity is rape – rape of nature, of the body politic, of vulnerable minorities, as much as that of women, both literal and metaphorical . . . This is happening to me and I feel violated. (Banuri 1994: 7)

> capital's history is aleatory because of its endless ability to *overcome* particular material conditions and obstacles, extending its reach ever further into all domains of discourse and, indeed, of existence . . . (Smith 1988: 141, emphasis mine)

In the globalization script, especially as it has been strengthened and consolidated since 1989, only capitalism has the ability to spread and invade. Capitalism is represented as inherently spatial and as naturally stronger than the forms of noncapitalist economy (traditional economies, "Third World" economies, socialist economies, communal experiments) because of its presumed capacity to univeralize the market for capitalist commodities. In its most recent guise, "the market" is joined by the operations of MNCs and finance capital in the irreversible process of spreading and spatializing capitalism. Globalization according to this script involves the *violation* and eventual death of "other" noncapitalist forms of economy:

> Capitalist production destroys the basis of commodity production in so far as the latter involves independent individual production and the exchange of commodities between owners or the exchange of equivalents. (Marx 1977: 951)

The globalization script normalizes an act of non-reciprocal penetration. Capitalist social and economic relations are scripted as penetrating "other" social and economic relations but not vice versa. (The penis can penetrate or invade a woman's body, but a woman cannot imprint, invade, or penetrate a Man.) Now that socialism is no longer perceived as a threat (having relinquished its toehold on strategic parts of the globe), globalization is the prerogative of capitalism alone. After the experience of penetration – by commodification, market incorporation, proletarianization, MNC invasion – something is lost, never to be regained. All forms of noncapitalism become damaged, violated, fallen, subordinated to capitalism. Negri's vision of "real subsumption" in the current phase of capitalist production and accumulation evokes the finality of this process:

> With the direct absorption by capital of all the conditions of production and reproduction, capital, through its command over the logic of social cooperation, envelops society and hence becomes social: capital has to

> extend its logic of command to cover the whole of society . . . (Note: It is not being claimed here that "real subsumption" precludes the existence of noncapitalist formations. Rather, in "real subsumption" a form of social cooperation exists that enables even noncapitalist formations to be inserted into a system of accumulation that renders them productive for capital.) (Surin 1994: 13, interpreting Negri)

To paraphrase Marcus (p. 399), the standardized and dominant globalization script constitutes noncapitalist economic relations as inevitably and only ever sites of potential invasion/envelopment/accumulation, sites that may be recalcitrant but are incapable of retaliation, sites in which cooperation in the act of rape is called for and ultimately obtained.

For the left, the question is, how might we challenge the dominant script of globalization and the victim role it ascribes to workers and communities in both "first" and "third" worlds? For, as Marcus suggests, to accept this script as a reality is to severely circumscribe the sorts of defensive and offensive actions that might be taken to realize economic development goals.

In connection with the dominant rape script, Marcus proposes two quite different courses of action. One is to change the script from within, to challenge it by refusing to accept the victim role. The other is to challenge the discourse of sexuality that the rape script inscribes and from which it draws its legitimacy and naturalness. Both challenges offer potential for thinking through alternative responses to globalization.

Becoming subject, unbecoming victim

> Rapists do not prevail simply because as men they are really, biologically, and unavoidably stronger than women. A rapist follows a social script and enacts conventional, gendered structures of feeling and action which seek to draw the rape target into a dialogue which is skewed against her. (Marcus 1992: 390)

To break from the victim role, Marcus suggests, a woman might change the script, use speech in unexpected ways, "resist self-defeating notions of polite feminine speech as well as develop physical self-defence tactics" (p. 389). She might remember that an erection is fragile and quite temporal, that men's testicles are certainly no stronger than women's knees, that "a rapist confronted with a wisecracking, scolding, and bossy woman may lose his grip on his power to rape; a rapist responded to with fear may feel his power consolidated" (Marcus 1992: 396).

Reading this I found myself asking, how might we get globalization to lose its erection – its ability to instill fear and thereby garner cooperation? I was drawn to an alternative reading of the globalization script that highlights contradictory representations of MNCs and their impacts upon "developing" economies.

The MNC[8] is seen as one of the primary agents of globalization in part because it provides an institutional framework for the rapid international mobility of capital in all its forms. Like the man of the rape script, the MNC is positioned in the standard globalization script as inherently strong and powerful (by virtue of its size and presumed lack of allegiance to any particular nation or labor force):

> TNCs are unencumbered with nationalist baggage. Their profit motives are unconcealed. They travel, communicate, and transfer people and plants, information and technology, money and resources globally. TNCs rationalize and execute the objectives of colonialism with greater efficiency and rationalism. (Miyoshi 1993: 748)

In "global factories" the MNC coordinates multiphase production processes in different countries and regions, seeking the cheapest combination of labor with appropriate skills to perform requisite tasks. But is this representation of a powerful agent able to assert and enact its will the only possible picture? Could we see the MNC in a different light – perhaps as a sometimes fragile entity, spread out and potentially vulnerable?

In the face of images of the all-pervasive global spread of MNC activity comes the contrary evidence that three of the four major *source* nations of transnational investment (the US, UK and West Germany, but not Japan) are also the major *hosts* of foreign direct investment (Dicken 1992: 87) and that "only perhaps 4 or 5 percent of the total population of TNCs in the world can be regarded as truly *global* corporations" (p. 49, emphasis in the original).[9] *Multi*national corporations, as Dicken points out, are more likely to be bi- or at the most quatrinational in their scale of global operation. And in light of the finding that most international productive investment flows between the so-called "developed" nations, the much publicized orientation of MNCs to areas of low wage costs bears re-examination. Most productive investment by US firms, for example, is market oriented, hence its

8 Also known as the TNC, or transnational corporation.

9 What is more, the magnitude of direct foreign investment (DFI) compared to domestic investment for these major investors is relatively small. Taking the example of the US, from 1960 to 1988 95.5 percent of total US investment was invested domestically. And if the eight nations that are the major sources of direct foreign investment are included, the ratio of DFI to total investment is less than 4 percent (Koechlin 1989, quoted in Graham 1993: 238).

focus upon Europe. Even investment in poorer countries may not be motivated by cost considerations but more by concern about access to final markets.[10] The representation of an all-encompassing global network of MNCs with a voracious appetite for cheap labor begins to crack in the light of these alternative illuminations, suggesting that the script of "capital flight" that dominates the political calculus of many activists in older industrial areas may be susceptible to revision.

The power of the MNC might also be open to discursive contestation. Corporate globalization is often constrained by language, law and business culture to certain national partnerships and not others. When it does reach beyond the safe channels between North America and Europe and selected countries around the Pacific Rim, it may be spread quite thin across national boundaries, vulnerable to being punched in the underbelly by changes in political leadership or trade policy or to being denied access (given the risk factor) to cheap venture capital. Increasingly, too, the MNC has become the prey of corporate raiders, the new knights of the global order. In the battle for ownership and control, size is no longer a protection. Small groups of individuals or relatively small firms organized into loose networks now lead takeover bids for much larger targets, and in most of these cases

> the "junk bond" financing arranged by these bidders necessitates that they liquidate a substantial portion of the target's assets to amortize these short-term bonds. (Coffee 1988: 116)

Almost overnight large MNCs may experience a change in control as shareholders exercise their "right" to mobilize and redesign their corporate possessions, forcing management to restructure and sometimes dismantle the company (Useem 1993).

If the spatial spread of many or even most MNCs encompasses only a handful of countries, then projects of "multinational" labor coordination might not be as difficult and daunting as they have sometimes appeared. Moreover, since institutional shareholders have begun to exercise their power over the operations of MNCs, there might be room for the institutional investment arms of the labor movement (pension and superannuation funds) to exercise their muscle in the corporate marketplace.

It would seem possible, therefore, to offer different (non-standard)

[10] Gordon estimates that much of the 25 percent of US foreign investment that does go to poorer countries is attempting to access markets that are closed by tariff and non-tariff barriers to foreign imports (1988: 50). And Tim Koechlin calculated that well over half the output of US MNCs in the Third World is sold locally (1991: personal communication).

responses to the set cues of the globalization script:

> And I grabbed him by the penis, I was trying to break it, and he was beating me all over the head with his fists, I mean, just as hard as he could. I couldn't let go. I was determined I was going to yank it out of the socket. And then he lost his erection . . . pushed me away and grabbed his coat and ran. (Bart and O'Brien, quoted by Marcus 1992: 400)

A non-standard response to the tactics of globalization was offered by the United Steel Workers of America (USWA) during the early 1990s when Ravenswood Aluminum Company (RAC) locked members of Local 5668 out over the negotiation of a new contract (Herod 1995). Research into the new ownership of RAC (which had been brought about by a leveraged buyout) established that the controlling interest was owned by a global commodities trader who had been indicted by the US Department of Justice on 65 counts of tax fraud and racketeering and was now residing in Switzerland to avoid his jail sentence (p. 350). The union surmised that the "key to settling the (contract) dispute . . . lay in identifying (the trader's) vulnerabilities and seeking to exploit them internationally" (p. 351). They decided (amongst other strategies) to allege that the company was a corporate outlaw "thereby potentially damaging its public image and encouraging government examination of its affairs" (pp. 351–2) and, through the channels of several truly international labor organizations, persuaded unions in a number of countries to lobby their own governments against dealing with such a corporate operator. Their interventions ended up ranging from Switzerland, Czechoslovakia, Rumania, Bulgaria, and Russia to Latin America and the Caribbean, all in all 28 countries on five continents. This international strategy – combined with a domestic end-user campaign to boycott RAC products – resulted (20 months after initiating the action) in the approval of a new three-year contract and in placing restrictions on the trader's expansion into Latin America and Eastern Europe.[11] Terrier-like, the USWA pursued the company relentlessly around the globe yanking and pulling at it until it capitulated. Using their own globalized networks, workers met internationalism with internationalism and eventually won.

If one way to challenge the rape script is to diminish the power of the perpetrator, another is to rescript the effects of the rape on the victim. Marcus suggests that feminists question the representations of rape as "theft," as "taking," as "death" – an event that is final and lasting

[11] Herod notes that the success of this dispute partly rested upon the financial resources that could be drawn upon from the International union and upon the ability to personalize the action – the trader made "an easy target for the locked out workers to vilify, which arguably strengthened their resolve to continue the corporate campaign against RAC" (1995: 354).

in terms of the damage it does to self.[12] How might we challenge the similar representation of globalization as capable of "taking" the life from noncapitalist sites, particularly the "Third World"? In the standard script the MNC conscripts Third World labor into its global factories, setting it to work in highly exploitative assembly plants, often located in export processing zones where production takes place for the world market, sealed off from the "local economy." The intervention of MNCs in the productive economy of the "host" country not only violates the "indigenous" economy, robbing it of its capacitity for self-generation, but also sterilizes it against future fecundity. In this sense it is a death ("We absorb the following paradox – that rape is death, but that in a rape the only way to avoid death is to accept it" [Marcus 1992: 395].) Yet is this representation the only one available? Could we not see MNC activity in Third World situations in a slightly different light, as perhaps sometimes unwittingly generative rather than merely destructive? In concocting this different story, there are a large number of industry studies to draw upon, of which studies of the semiconductor industry are perhaps the most numerous and compelling.

The semiconductor industry has become a well-known instance of the global factory. The most famous product of this industry is the integrated circuit, which is produced in a multiphase process that is often spatially dispersed. Assembly work, or the application of microscopic circuitry to silicon chips, is frequently performed in Asia by young women at low wages. The assembled chips are then re-imported to the "home" country where they are used in the production of electronic products like computers and computerized equipment. Labor relations and working conditions in the export processing zones where chips are assembled have often been described as repressive and barbaric.[13] But it is important that we not accept this model of economic sterility and social violence as the "inevitable" outcome of MNC operations in the Third World.

Recent developments in the international semiconductor industry indicate that the penetration of Asia by foreign MNCs has borne unexpected fruit. Both upstream suppliers and downstream users of semiconductors

[12] "The rape script strives to put women in the place of objects; property metaphors of rape similarly see female sexuality as a circumscribable thing. The theft metaphor makes rape mirror a simplified model of castration: a single sexual organ identifies the self, that organ is conceived of as an object that can be taken or lost, and such a loss dissolves the self. These castration and theft metaphors reify rape as an irrevocable appropriation of female sexuality" (Marcus 1992: 398).

[13] Spivak (1988b: 89), for example, draws attention to the way in which the government, male workers and company management conspired to violently quash union activity among women workers in a South Korean semiconductor plant owned by Control Data.

have sprung up in a regional complex of indigenous firms in south-east Asia, including technical training facilities, centered in Thailand (Scott 1987). This development counters the image of the sterile branch plant in poor countries, which repatriates profits and contributes only to underdevelopment rather than industrial growth. The development of the locally-owned semiconductor industry in south-east Asia has been extremely dynamic and relies increasingly on highly skilled local workers.[14] It appears that the economic "rape" wrought by globalization in the Third World is a script with many different outcomes. In this case we might read the rape event as inducing a pregnancy, rather than initiating the destruction and death of indigenous economic capacity.

Of course, one might be concerned that penetration by the MNCs has instigated the reproduction of many more rapists, indigenous capitalists eager to enact their own sinister scripts. But there are other indications that the effects of MNC capitalist activity in the Third World has had implications for noncapitalist activity as well. Recent feminist research has emphasized the dynamic and conflictual interactions between capitalist factory employment, households, the state, and "Third World women," supplanting the mechanistic narrative of women's subsumption to the logic of capital accumulation with multiple stories of complexity and contestation (Elson and Pearson 1981; Pearson 1986; Ong 1987; Kondo 1990; Lim 1990; Cameron 1991; Wolf 1992; Mohanty et al. 1991).[15]

The proletarianization wrought by globalization has created, for many Third World women, a more complex social formation, one which is not dominantly or only capitalist. Participation in the capitalist class process of surplus value production and appropriation in global production

[14] Scott (1987) found local producers engaged in all phases of the production process from design to testing, though assembly subcontracting was the major activity at the time. To challenge the globalization script and point out the local "vitality" of the industry is not, however, to deny or dismiss the importance of the conditions under which many women labor in these firms.

[15] The standard globalization script highlights the employment of cheap, docile and compliant female wage labor in Asia and Latin America by industrial corporations eager to escape the expensive, organized, and predominantly male labor forces of the industrial capitalist world (Armstrong and McGee 1985; Frobel et al. 1980). Actually, though, the industries that use female labor in the Third World are usually the ones that employ women in their home countries, such as textiles, garments and electronics (Porpora et al. 1989). In the standard script the patriarchal household has a non-contradictory relation to capitalist exploitation, facilitating and deepening it. Patriarchy plays the support role to capitalism, enforcing women's participation in wage labor, as families drive their daughters into the labor market in order to get access to a wage that will be dutifully remitted by them. It also enhances exploitation, as employers take advantage of the devaluation of women's labor and the female docility that is patriarchally induced.

sites has interacted in a variety of ways with women's participation in class processes in their households.[16] For some women, involvement in capitalist exploitation has freed them from aspects of the exploitation associated with their household class positions and has given them a position from which to struggle with and redefine traditional gender roles. Strauch (1984), for example, writes of the abandonment by women factory workers from rural Chinese families in Malaysia and Hong Kong of the traditional patriarchal custom of living with their husbands' families and of their success in establishing independent households or moving conjugal families back to their childhood villages or homes. Urban employment has unsettled traditional domestic class practices in which their surplus labor was destined to be appropriated by their husbands' families.[17]

In addition to the changed insertion of women into the traditional household, the emergence of factory employment has been associated in certain households with the rise of non-traditional domestic class processes. Women are now often the only income earners in their households and may live on their own as single women or as single mothers whose husbands have abandoned the family or migrated in search of work (Cameron 1991; Heyzer 1989). With no male head of household present to appropriate their surplus labor, these women are engaged in the independent domestic class process of producing, appropriating, and distributing their own surplus labor. Their example, while still relatively uncommon, may encourage other women to participate in household struggles for class transformation, leaving their husbands' families or even their husbands to establish new households along independent class lines. Communal households in which household members jointly produce and appropriate their surplus labor also may be emerging as one of the consequences of factory employment options for Third World women.[18] Globalization can be seen, then, as overdetermining the emergence of different noncapitalist class processes in the household.[19]

Reading globalization as Marcus reads rape, as a scripted series of steps and signals, allows me to see the MNC attempting to place regions, workforces and governments in positions of passivity and victimization and being met by a range of responses – some of which play into

[16] See chapter 3 for an introduction to a conceptualization of class and class processes.

[17] Under their new circumstances, they were expected to contribute labor to their natal families if proximity allowed; but the degree of exploitation associated with these new transfers of labor – though not explicitly assessed by Strauch – was apparently much less than that experienced by the traditional daughter-in-law, both because the transfer of surplus labor was not enforced by longstanding tradition and because of other householding options available to the women themselves.

the standard script and others that don't. It helps me to challenge the hegemonic representation of the superior power of the MNC by seeing how the conditions of existence of that power are constituted in language as much as in action, and even more importantly, in a complex interaction between the two. As Ernesto Laclau has argued, we do ourselves an injury, and promote the possibility of greater injury, by accepting a vision in which "absolute power has been transferred . . . to the multinational corporations":

> A break must be made with the simplistic vision of an ultimate, conclusive instance of power. The myth of liberal capitalism was that of a totally self-regulating market from which state intervention was completely absent. The myth of organized capitalism was that of a regulatory instance whose power was disproportionately excessive and led to all kinds of wild expectations. And now we run the risk of creating a new myth: that of the monopoly corporations' limitless capacity for decision-making. There is an obvious symmetry in all three cases: one instance – be it the immanent laws of the economy, the state or monopoly power – is presented as if it did not have conditions of existence, as if it did not have a constitutive outside. The power of this instance does not therefore need to be hegemonically and pragmatically constituted since it has the character of a ground. (Laclau 1990: 58–9)

His belief that he has more strength than a woman and that he can use it to rape her merits more analysis than the putative fact of that strength,

[18] Whereas the state is often theorized as promoting capital accumulation by "liberating women's labour to capital" (Pearson 1986: 93), it is less often seen as overdetermining a class transition in the household. In Singapore, however, where there was a shortage of labor, women's labor was seen as crucial for an export-oriented development strategy. As industrialization has progressed, the state has encouraged men to play a greater role in child-care and housework in order to make it possible for more women to work outside the home (Phongpaichit 1988). The Singaporean state can thus be seen not only as facilitating the extension of capitalist class processes but as fostering the development of communal domestic class processes as well.

[19] Even in cases where the class nature of the traditional household is unchanged, women's exploitation at home may stand in contradictory relation to their exploitation in the factory. Rather than merely constituting a woman as less militant and more exploitable in the capitalist workplace, the woman's double day may also be a condition of her activism. In a certain Thai factory, for example, the women were not able to take advantage of management invitations to socialize after work. They developed a view of these social events as a "management ploy" and were much less likely to be co-opted by management than were their male coworkers (Porpora et al. 1989). Over time, their suspicions of management developed into a more organized and ultimately successful struggle for improved working conditions. Ironically, then, for the married women at this factory, successfully changing their conditions of capitalist exploitation was an overdetermined and contradictory effect of their continued exploitation in a feudal domestic class process at home (Cameron 1991: 22).

> because that belief often produces as an effect the male power that appears to be rape's cause. (Marcus 1992: 390)

Inscribing sexual/economic identity

Rewriting the script of gendered violence so that women are no longer subjects of fear and objects of violence but become subjects of violence (and therefore capable of real opposition) in their own right is enabled by discussions and evidence that challenge the unquestioned capability of male bodies to rape, and that question their superior strength and the impossibility of resistance. Stories of resistance, of cases where women averted rape or fought back, provide empowering images that might help women to "rewrite" the standard rape script. But Marcus is concerned to go beyond the tactics of reversal and individual empowerment. This leads her to another kind of interest in the metaphor of a script.

One of the powerful things about rape in our culture is that it represents an important *inscription* of female sexual identity. Marcus argues that we should

> view rape not as the invasion of female inner space, but as the forced *creation* of female sexuality as a violated inner space. The horror of rape is not that it steals something from us but that it *makes* us into things to be taken . . . The most deep-rooted upheaval of rape culture would revise the idea of female sexuality as an object, as property, and as inner space. (Marcus 1992: 399 emphasis mine)[20]

The rape act draws its legitimacy (and therefore the illegitimacy of resistance) from a very powerful discourse about the female body and female and male sexuality. Thus the subject position of victim for the woman is not created merely by the strength and violence of the rapist but also by the discourse of female sexual identity that the rape script draws upon in its enactment:

> Rape *engenders* a sexualized female body defined as a wound, a body excluded from subject–subject violence, from the ability to engage in a fair fight. Rapists do not beat women at the game of violence, but aim to exclude us from playing it altogether. (Marcus 1992: 397 emphasis mine)

In Marcus's view, the creation of alternatives to the standard rape script is predicated upon a significant revision of the very powerful discourses

[20] Marcus continues: "Thus, to demand rights to ourselves as property and to request protection for our vulnerable inner space is not enough. We do not need to defend our 'real' bodies from invasion but to rework this elaboration of our bodies altogether" (1992: 399).

of sexual identity that constitute the enabling background of all rape events:

> New cultural productions and reinscriptions of our bodies and our geographies can help us begin to revise the grammar of violence and to represent ourselves in militant new ways. (Marcus 1992: 400)

The feminist project of rewriting and reinscribing the female body and sexuality has born much theoretical fruit in recent years, prompted especially by the work of French philosophers Irigaray, Kristeva, and Cixous. Marcus's challenge and encouragement to marry this rethinking with strategizing a new (poststructuralist) feminist politics of rape prevention has prompted me to further thoughts about globalization and the politics of economic transformation. In particular, they have led me to consider how a rewriting of the male body and sexuality might affect views of capitalism and its globalizing capacities:

> Men still have everything to say about their sexuality, and everything to write. For what they have said so far, for the most part, stems from the opposition activity/passivity, from the power relation between a fantasized obligatory virility meant to invade, to colonize, and the consequential phantasm of woman as a "dark continent" to penetrate and "pacify." (Cixous 1980: 247)

Feminist theorists have generated many new representations to replace that of the vacant, dark continent of female sexuality. However, as Grosz remarks, the particularities of the male body that might prompt us to challenge its naturalized hard and impermeable qualities have largely remained unanalyzed and unrepresented (1994b: 198). Discussion of bodily fluids, for example, is rarely allowed to break down the solidity and boundedness of the male body:

> Seminal fluid is understood primarily as what it makes, what it achieves, a causal agent and thus a thing, a solid: its fluidity, its potential seepage, the element in it that is uncontrollable, its spread, its formlessness, is perpetually displaced in discourse onto its properties, its capacity to fertilize, to father, to produce an object. (Grosz 1994b: 199)

Grosz's suggestive words offer a brief glimpse of how we might differently conceive of the body of capitalism, viewing it as open, as penetrable, as weeping or draining away instead of as hard and contained, penetrating, and inevitably overpowering. Consider the seminal fluid of capitalism – finance capital (or money) – which has more traditionally been represented as the lifeblood of the economic system whose free circulation ensures health and growth of the capitalist body (see chapter 5). As seminal fluid, however, it periodically breaks its bounds, unleashing uncontrollable gushes of capital that flow every which way, including into self-destruction. One such spectacle of bodily excess, a wet dream that stained markets around the globe, occurred in October 1987, when stock markets across the world crashed, vaporizing millions of dollars in immaterial wealth (Wark 1994: 169). The 1987 crash was for many the bursting of a bubble of irrationality, "a suitable ending, fit for a moral fable, where the speculators finally got their just deserts" (Wark 1994: 171–2). But the growth of activity on international financial markets represents an interestingly contradictory aspect of the globalization script.

One of the key features of globalization has been the complete reorganization of the global financial system since the mid-1970s:

> The formation of a global stock market, of global commodity (even debt) futures markets, of currency and interest rate swaps, together with an accelerated geographical mobility of funds, meant, for the first time, the formation of a single world market for money and credit supply. (Harvey 1989: 161)

On the one hand this growth has been seen to facilitate the rapid internationalization of capitalist production, and the consolidation of the power of finance capital is seen to represent the "supreme and most abstract expression" of capital (Hilferding, quoted in Daly, 1991: 84). On the other, this growth has unleashed money from its role as a means of circulation and allowed the rampant proliferation of global credit:

> Now detached from its necessity to abide by the laws of capital accumulation the finance sector oversees and facilitates capital movement into speculation, mergers, acquisitions – "financial gamesmanship." (Harrison and Bluestone 1988: 54)

Money has become

> a kind of free-floating signifier detached from the real processes to which it once referred. Through options, swaps and futures, money is traded for money. Indeed since much of what is exchanged as commodities are future monetary transactions, so what is traded in no sense exists. (Lash and Urry 1994: 292)

The globalization of credit and financial markets has created an "opening" for capital to transcend its own limits:

> The economy is an ensemble of movements and flows, mostly tied more or less to the physical space of fixed assets that persist in time. The financial vector is a dynamic development that seeks to escape from commitment to such permanence. (Wark 1994: 176)[21]

For those interested in historical periodizations of capitalism, this development signifies the demise of capitalism in its "second nature"

(Debord 1983), when the productive base anchored the movements of credit and money.[22] Today, the whole relation between signifier, signified and referent has been ruptured thereby unleashing capitalism's "third nature," "the spectacle" (Debord 1983), "the enchanted world" (Lipietz 1985) in which the "economic real" is buried under the trade of risk, information, image, futures, to be revealed only in momentary displays such as occurred when stock markets crashed in October 1987 (Wark 1994). At such orgasmic moments the body of capitalism undergoes a spasm of uncontrollability and unboundedness, to be brought back to itself and recontained within its own limits during the little death that follows:

> Capital pushes the development of the [finance] vector so hard and fast that it bursts through its own limit, as it did in October 1987. Like a meltdown, the crash was an accident programmed in advance to happen as the system pushed against its own limits . . . In attempting to increase its power over itself and the world via the vector, capital encounters obstacles within itself . . . Third nature is both the means by which capital extends itself, and the symptom of its inability to do so . . . The event [the crash] is a privileged moment in which to see third nature for what it really is, stripped of its myths and kitchen gods. (Wark 1994: 192)

Globalization, it seems, has set money free of the "real economy" and allowed capital to seep if not spurt from the productive system, but the implications of this unboundedness, this fluidity, for the identity of capitalism remains unexplored. Having set the signifier free from the referent, theorists of the global economy are loath to think about the effects of seepage, porosity, uncontrollability, that is, to feminize economic identity (Grosz 1994b: 203). The global economy may have been opened up by international financial markets, but nothing "other" comes into or out of this opening. It would seem that the homophobia that pervades economic theorizing places a taboo on such thinking:[23]

> Part of the process of phallicizing the male body, of subordinating the rest

[21] Many have drawn attention to the detrimental effects such developments have upon the productive "base." Share trading can dissolve corporations over night, productive investments can be milked for cash to play the currency markets and managers of corporations can be asked to trade themselves out of a job (Coffee, Lowenstein, and Rose-Ackerman 1988). Mitchell (1995: 369) refers to the credit system characteristic of "casino capitalism" as a "feral credit system" running wild and uncontrolled, which "seems completely disconnected from the productive economy" in which the movement of goods has become "more and more negotiated and regulated" throughout the 1980s (Thrift 1990: 1136).

[22] That is, capitalism as we knew it under the names of, for example, "organized capitalism" (Lash and Urry 1987) or Fordism.

> of the body to the valorized functioning of the penis, with the culmination of sexual activities occurring ideally at least, in sexual penetration and male orgasm, involves the constitution of the sealed-up, impermeable body. Perhaps it is not after all flow in itself that a certain phallicized masculinity abhors but the idea that flow moves or can move in two-way or indeterminable directions that elicits horror, the possibility of being not only an active agent in the transmission of flow but also a passive receptacle. (Grosz 1994b: 200–1)

How might we confront the economic "unthinkable," engendering a vision of the global economy as penetrable by noncapitalist economic forms? Perhaps the very same financial system that is represented as both the agent of seepage and the agent of capital's assertion of identity might yield a further surplus of effects. Perhaps we might see the proliferation of credit and deregulation of financial markets as creating opportunities for the growth of noncapitalist class relations as well as capitalist ones. The huge expansion of consumer credit (including credit card financing with large maximum limits, home equity loans, and a variety of other instruments almost forced upon "consumers") is often assumed to promote personal indebtedness associated with a culture of consumption. Yet, given the growth in self-employment and of home-based industries – some of which is associated with the downsizing and streamlining of capitalist firms – it is clear that much of what is seen as consumer credit is actually (or also) producer credit, in other words it is used to buy means of production (including computers and other equipment) and other inputs into the production process of self-employed workers. Historically such loans have been notoriously difficult to obtain from traditional financial institutions like local and regional banks, but with the growth of new international credit markets they have become quite instantaneous and straightforward. This has contributed to an increase in small businesses that are sites of noncapitalist class processes of individual and collective surplus appropriation (as well as providing a source of credit to small capitalist firms). The financial sector can be seen, then, as an *opening* in the body of capitalism, one that not only allows capital to seep out but that enables noncapitalism to invade.

The script of globalization need not draw solely upon an image of the body of capitalism as hard, thrusting and powerful. Other images are

[23] Indeed, as Whitford suggests, "the house of the male subject is closed . . . whereas one characteristic of woman's sexual bodies is that they are precisely not closed; they can be entered in the act of love, and when one is born one leaves them, passes across the threshold. (One might argue that men's bodies can be entered too, but no doubt Irigaray would argue that the massive cultural taboo on homosexuality is linked to men's fear of the open or penetrable body/houses)" (1991: 159).

available, and while we cannot expect the champions of globalization to express pleasure in leakage, unboundedness and invasion, it is important to draw upon such representations in creating an anticapitalist imaginary and fashioning a politics of economic transformation. If the identity of capitalism is fluid, able to penetrate and be penetrated, then the process of globalization need not constitute or inscribe "economic development" as inevitably *capitalist* development. Globalization might be seen as liberating a variety of different economic development paths. In fact, the script of globalization may already (without explicit instances of opposition) be engendering economic differences.[24]

Marcus encourages a rejection of the fixed sexual identity that is *inscribed* upon the body of women by the rape script. Given that this identity is rooted in a dominant discourse of heterosexuality (in which the bodies of men and women are distinguished by rigid and fixed gender differences and in which male and female behavior can only be understood in terms of opposition, complementarity or supplementarity), one of the implications of Marcus's argument is that rape (along with marriage) is a recognized and accepted (if not acceptable) practice of heterosexuality. Challenging the legitimacy our culture implicitly grants to rape becomes, in this formulation, a challenge to heteronormativity itself.

My desire to reject globalization as the inevitable inscription of capitalism prompts me to take Marcus's implication one step further and to explore the ways in which discourses of homosexuality might liberate alternative scripts or inscriptions of sexual/economic identity:

> many gay men . . . are prepared not only to send out but also to receive flow and in this process to assert other bodily regions than those singled out by the phallic function. A body that is permeable, that transmits in a circuit, that opens itself up rather than seals itself off, that is prepared to respond as well as to initiate, that does not revile its masculinity . . . or virilize it . . . would involve a quite radical rethinking of male sexual morphology. (Grosz 1994b: 201)

Queering globalization/"the universality of intercourse"

Rethinking capitalist morphology in order to liberate economic development from the hegemonic grasp of capitalist identity is indeed a

[24] There are other ways of conceiving of capitalism as open, as invadable or even as invaded. Theorizing the household as a site of noncapitalism, as a site, that is, of many noncapitalist class processes, generates a vision of the putatively hard, impenetrable and capitalist body economic as always carrying within it economic identities that are noncapitalist.

radical project. Yet resources for such a project are already available in the domain of social theory, especially within queer theory, where a rethinking of sexual morphology is taking place. For queer theorists, sexual identity is not automatically derived from certain organs or practices or genders but is instead a space of transitivity (Sedgwick 1993: xii):

> one of the things that "queer" can refer to [is] the open mesh of possibilities, gaps, overlaps, dissonances and resonances, lapses and excesses of meaning when the constituent elements of anyone's gender, of anyone's sexuality aren't made (or *can't be* made) to signify monolithically. (Sedgwick 1993: 8, emphasis in original)

Sedgwick's evocation of the way in which things are supposed to "come together" at certain social sites – she speaks of the family, for example, where a surname, a building, a legal entity, blood relationships, a unit in a community of worship, a system of companionship and caring, a sexual dyad, the prime site of economic and cultural consumption, and a mechanism to produce, care for, and acculturate children[25] are "meant to line up perfectly with each other" (1993: 6) – captures the oppressiveness of familiar identity-constituting/policing discourses and provides a glimpse of the productiveness of fracturing and highlighting dissonances in seemingly univocal formations. Through its challenges to such consolidities of correspondences and alignments, queer theory has encouraged me to attempt to rupture monolithic representations of capitalism and capitalist social formations.

One key conflation or "coming together" that participates in constituting a capitalist monolith is the familiar association of capitalism with "commodification" and "the market." When it is problematized, which is not very often, the presumed overlap between markets, commodities and capitalism will often be understood as the product of a capitalist tendency to foster what Marx calls "the universality of intercourse" (1973: 540). This tendency is central to the representation of the capitalist body as inherently capable of invading, appropriating and destroying:

> while capital must on one side strive to tear down every spatial barrier to intercourse, i.e. to exchange, and conquer the whole earth for its market, it strives on the other side to annihilate this space with time, i.e. to reduce to a minimum the time spent in motion from one place to another . . . There appears here the universalizing tendency of capital, which distinguishes it from all previous stages of production. (Marx 1973: 539–40)

When the "market" is invoked today, nearly fifty years after the construction of the post-war order, the main referent must be to the world market

[25] All these features and more are listed in Sedgwick (1993: 6).

> – to the markets for goods and services, capital (or "loci of production"), money and credit. The world market is the site of economic reproduction of the global capital relation, as well as of the political organization of hegemony. An opening to the world market is thus synonymous with integration into the global process of economic reproduction and a historically determined system of hegemony. (Altvater 1993: 80–1)[26]

In discussions of the bodily interactions (that is, market transactions) between capitalism and its "others," the metaphor of infection sometimes joins those of invasion and penetration:

> capital is the virus of abstraction. It enters into any and every social relation, corrupts it, and makes it manufacture more relations of abstraction. It is a form of viral relations which has a double aspect. It turns every qualitative and particular relation into a quantitative and universal one. (Wark 1994: xii)

It is perhaps not surprising that capitalism is represented as a body that invades and infects, but is not itself susceptible to invasion or infection. In the shadow of the AIDs crisis queer theorists have shed light upon the homophobia that pervades social theory (if not the specific heteronormativity of economic theory) (Sedgwick 1993). The infection metaphor suggests an "immensely productive incoherence" (to use Sedgwick's phrase [1993: xii]) that may help to disrupt the seamlessness of capitalist identity, reconfiguring capitalism's morphology in ways that our earlier rethinking of heterosexual rape and penetration could not.

The locus and agent of universalizing intercourse is "the market," which continually seeks new arenas/bodies in which to establish a medium, or circuitry, through which contamination by capitalism may flow. Globalization discourse highlights the one-way nature of this contamination and the virtual impossibility of immunity to infection. But markets/circuits cannot control what or who flows through them. The market can, in fact, communicate many diseases, only one of which is capitalist development. Consider as an example the increasingly international labor market, another character in the standard script of globalization. The huge increase in international migration and the establishment of large immigrant "underclasses" in the metropolitan capitals (world cities) of the global economy since the 1960s is attributed both to economic and political destabilization in "donor" countries and to the voracious appetite of "first world" capitalism for cheap labor, particularly in the growing service and low-wage manufacturing industries

[26] By contrast, Daly notes: "There is nothing . . . which is essentially capitalist about the market. Market mechanisms pre-exist capitalism and are clearly in operation in the socialist formations of today. It is clear, moreover, that a whole range of radical enterprises exist within the sphere of the market . . . " (1991: 88).

(Sassen 1988). The loss of labor from source countries is usually portrayed as yet another outcome of the penetration of capitalism, via a process in which "the market" or "commodification" has destroyed the traditional, often agricultural, economy and forced people into proletarianization.

When incorporated into a script of globalization, the extraordinary economic diversity of immigrant economies in "host" countries is subsumed to, or depicted as an "enclave" within, the capitalist totality, located on a trajectory of homogenization or synthesis with the host society. The noncapitalist nature of immigrant entrepreneurial activity is documented for its cultural interest, but is rarely allowed the autonomy afforded to "capitalist enterprise." Yet immigrant economies made up of self-employed and communal family-based enterprises (as well as small capitalist enterprises) operate their own labor and capital markets, often on a global scale (Collins et al. 1995; Waldinger 1986). They maintain complex economic as well as political and cultural connections with other diasporic communities and with their "home" countries.

Even immigrant workers who may be wage laborers in metropolitan capitalist economies cannot be seen as "subsumed" in any complete sense. Rouse (1991) documents how Mexican immigrants working as wage laborers in the service industries of US cities manage two distinct ways of life, maintaining small-scale family based farms or commercial operations in Mexico. Money received as wages in the US is siphoned into productive investment in noncapitalist activity in the Mexican economy:

> Aguilillans have come to link proletarian labor with a sustained attachment to the creation of small-scale, family based operations . . . Obliged to live within a transnational space and to make a living by combining quite different forms of class experience, Aguilillans have become skilled exponents of a cultural bifocality that defies reduction to a singular order. (Rouse 1991: 14–15)

Globalization, it would seem, has not merely created the circuitry for an increased density of international *capital* flows: "Just as capitalists have responded to the new forms of economic internationalism by establishing transnational corporations, so workers have responded by creating transnational circuits" (Rouse 1991: 14). Labor flows have also grown, and with them a variety of noncapitalist relations.

This case is a wonderful example of the productive incoherence that can be generated through a metaphor of infection – an infection whose agent is the market (in this instance the international labor market). It challenges the imperial nature of capitalism, depriving capitalism of its role as sole initiator of a spatially and socially expansive economic circuitry of infection. If capitalist globalization is an infection, it can be said to coexist with many other types of infection. Were we not bedazzled by images of the superior morphology of global capitalism, it might be possible to theorize the global integration of noncapitalist economic relations and non-economic relations and to see capitalist globalization as coexisting with, and even facilitating, the renewed viability of noncapitalist globalization.

Another productive incoherence introduced by the metaphor of infection arises from the possibility of immunity. Lash and Urry note that "goods, labour, money and information will not flow to where there are no markets" such as into eastern Europe after the devastation of state economic governance or into the "black ghetto in the USA" (1994: 18). The globalization of the capitalist market is a spotty affair – characterized by cobweb-like networks of density and sparseness (p. 24), areas of infection and areas of immunity. Within the areas of sparseness/immunity the economic transactions that take place are seen to be of no consequence, and yet it is easy to imagine that in the interstices of the (capitalist) market, other markets do exist. And in these other markets noncapitalist commodities are exchanged.

But the market is notoriously catholic in its tastes and desires for the exotic, the noncapitalist, the unclean. And, so goes the standard script, sooner or later, these areas of immunity, of capitalist vacuity, of noncapitalist commodity production, will come under the power of capitalism. It is only a matter of time before capitalism reaches out via "the market" into the unknown, infecting subjects with the desire for capitalist commodities and gathering to itself objects for exchange:

> Capitalist culture industries typically seek out the "new" in the cultures of various "others" . . . African American, Third World, gay and lesbian, working-class, and women's (sub)cultures frequently serve the function of research and development for cultural capitalists. (MacNeill and Burczak 1991: 122)[27]

> it is not the people who have nothing to lose but their chains, but the flow of qualitative information about people, their places, and the things they produce, in short their "culture." Culture . . . ceases in the process to be culture, and becomes instead a postculture or a transculture . . . Culture is something that will be overcome – whether we like it or not. (Wark 1994: xiii)

[27] Businesses that anchor their identities upon the appropriation of "Third World" or "tribal" artifacts and knowledges are good examples of this trend: "The tribal world, it turns out, is full of traditional wisdom, especially in providing recipes for products that can be sold at The Body Shop. The urge to 'globalize' a society for its own purposes may not be entirely a new phenomena (sic) in the corporate world. However, in the case of The Body Shop, it is being done as a practice integral to the very identity of that corporation" (Deshpande and Kurtz 1994: 47).

Contact, via the market, with noncapitalist commodities or "other cultures" never permits the transmission of infection back into the capitalist body. Instead, in the wake of the capitalist market comes the commodification of culture and the death of "authentic culture."

The homogenizing claims made for the global market and capitalist commodification are, in the eyes of Grewal and Kaplan, insufficiently interrogated:

> What is not clear in such debates is what elements of which culture (including goods and services) are deployed where, by whom, and for what reason. Why, for instance, is the Barbie doll sold in India but not the Cabbage Patch doll? Why did the Indian Mattel affiliate choose to market Barbie dressed in a sari while Ken remains dressed in "American" clothes? . . . Which class buys these dolls and how do children play with them in different locations? . . . Getting away from . . . "transnational centrism" acknowledges that local subjects are not "passive receptacles" who mechanically reproduce the "norms, values, and signs of transnational power." (1994: 13)

The market is not all or only capitalist, commodities are not all or only products of capitalism, and the sale of Barbie dolls to Indian girls or boys does not at all or only presage the coming of the global heterosexist capitalist kingdom. A queer perspective can help to unsettle the consonances and coherences of the narrative of global commodification. That there need not be a universal script for this process is suggested by Mary Gossy's discussion of the use to which a Barbie-like doll is put in a series of photographs entitled "Gals and Dolls":[28]

> Barbie is not a dildo, not a penis substitute, but rather a figure of lesbian eroticism, a woman in the palm of a woman's hand. In fact, playing with Barbies and Kens teaches girls lesbian butch/femme roles – and thus in puberty the doll becomes a representative of lesbian desire. (1994: 23)

In the same spirit, might we not wonder exactly what is being globalized and how infection is moving when Barbie dolls are marketed in India or, for that matter, Indiana?[29] What power and potentialities do we relin-

[28] The photographs show a Barbie-like doll emerging from or entering a vulva.

[29] Rand concludes her fascinating queer exploration of Barbie memories and practices within the US with this reflection: "Surprisingly often, the stories I heard were about how Barbie turned people into cultural critics and political activists – about how seeing activists queer Barbie, or remembering their own Barbie queerings, or hearing about my Barbie work, induced them to move from Barbie anecdotes to thinking about cultural politics, ideology, oppression, and resistance, and sometimes to political plotting and practice" (1995: 195).

quish when we accept the univocality of the market/commodity/global capitalist totality?[30]

If we create a hegemonic globalization script with the MNC, the financial sector, the market and commodification all set up in relations of mutual reinforcement, and we then proclaim this formation as a "reality," we invite particular outcomes. Certain cues and responses will be seen as "normal" while others will be seen as quixotic and unrealistic. By querying globalization and queering the body of capitalism we may open up the space for many different scripts and invite many different actors to participate in the realization of different outcomes.

Conclusion: Globalization and its "other" or globalization as "other"

> To treat rape simply as one of . . . "the realities that circumscribe women's lives" can mean to consider rape as terrifyingly unnameable and unrepresentable, a reality that lies beyond our grasp and which we can only experience as grasping and encircling us. (Marcus 1992: 387)

The discourse of globalization that I have been discussing in this chapter presents capitalism as something that is certainly "grasping and encircling us." Against the force of globalization many have focused upon the "local" (or localization) as its only "other." Localization invokes the way in which "global processes can in a sense be pinned down in certain localities and hence can become the basis for self-sustaining growth in those places" (Lash and Urry 1994: 284). "As spatial barriers diminish so we become much more sensitized to what the world's spaces contain" (Harvey 1989: 294). Theorists such as Haraway have been concerned to make visible the many and heterogeneous globalized organizations that build alliances across situatednesses and establish "webs of systematicity" between locales.

But while localization may involve certain resistances to global processes, there is no room for the penetration of globalization/capitalism by the local. Localization, it seems, is not so much "other" to globalization as contained within it, brought into being by it, indeed part of globalization itself.

Marcus's discussion of the rape script has prompted me to explore

30 Daly points to some other ways of destabilizing the capitalist totality: "there has been a corresponding shift towards more politically aware forms of lifestyle consumerism – emphasis on local produce, environmentally sound products, anti-apartheid, anti-vivisectionist goods etc. And these new developments cannot be thought within the classical identification of the new economic space as a fixed capitalist totality, or as a homogeneous milieu . . . " (1991: 88).

the ways in which we might resist and rethink the representation of globalization as the social disciplinarian that polices all economic transactions – forcing them into line, into direct competition and equilibration – and thereby establishes a Kingdom on earth in which the local is humbled.[31] Her rethinking of rape has inspired me to imagine the repositioning of subjects in the globalization script, and to think differently about globalization (based upon a differently constituted morphology of capitalist identity) as itself "other."

The strategies suggested by Marcus involve first rewriting the globalization script from within, denying the inevitability and "reality" of MNC power over workers and communities and exploring ways in which the hard and penetrating body of the MNC can be seen as soft, fragile, and vulnerable. Making global capitalism lose its erection becomes a real possibility if we reject the naturalization of power and violence that is conferred upon the MNC by the globalization script. It also becomes possible to challenge the representation of economic penetration as necessarily inducing sterility or causing death in the Third World body economic.[32] Both these rewritings attempt to generate a vision of alternative scripts and outcomes for economic transformation without challenging the naturalness of (capitalist) globalization itself.

The second strategy for querying globalization entails identifying the larger discourse of economic identity and development that grants "reality" and legitimacy to global capitalism, and exploring the inscription by the globalization script of a hegemonic capitalist identity upon the world body economic. Drawing upon feminist retheorizations of sexual identity, the naturalness of capitalist identity as the template of all economic identity can be called into question. We may attempt to make globalization less genital, less phallic, by highlighting various points of excess in its inscriptions – places where the inscription can be seen as uncontrollable or indeterminate, or as potentially inscribing noncapitalist identity.

Finally, the severing of globalization from a fixed capitalist identity is enabled by a queering of economic identity, a breaking apart of the monolithic significations of capitalism (market/commodity/capital) and a liberation of different economic beings and practices. A space can be made for thinking globalization as many, as other to itself, as inscribing different development paths and economic identities. Globalization need

[31] "It is curious how the market seems to be the central metaphorical term for equal but opposite narratives of legitimation. On the one hand, the market means freedom; on the other, it means discipline" (Wark 1994: 181).

[32] This representation draws its force from the conceptualization of national economies as self-contained organisms that compete for survival. See chapter 5 for a discussion of the economy as a body or organism.

not be resisted only through recourse to the local (its other within) but may be redefined discursively, in a process that makes room for a host of alternative scriptings, capable of inscribing a proliferation of economic differences.

> Rape does not happen to preconstituted victims; it momentarily makes victims. The rapist does not simply *have* the power to rape; the social script and the extent to which that script succeeds in soliciting its target's participation help to create the rapist's power. The rape script pre-exists instances of rape but neither the script nor the rape act results from or creates immutable identities of rapist and raped. (Marcus 1992: 391)

7

Post-Fordism as Politics

Reading the growing literature on post-Fordism, I am impressed by the power of this "model of development" to organize and illuminate contemporary experience, interpreting and connecting a wide variety of social processes and institutions.[1] Yet at the same time that I have been impressed by the fertility and richness of the literature that has amassed around this catalytic concept, I have found aspects of post-Fordist theory quite problematic and constraining. By emphasizing the thoroughly capitalist nature of industrial social formations, by theorizing societies as centered upon economies, by representing contradiction as mediated or stalled, and by understanding development as a systemic or hegemonic process, many theorists of post-Fordism have replicated the characteristics of other and earlier theories of (capitalist) development.

For those who use such theoretical work as a guide to policy proposals and political action, such a continuity with earlier Marxist ways of understanding industrial societies has an insidious edge. Once again a stable, coherent, and hegemonic formation has been placed in the path of the forces of change.

Yet what post-Fordist theory takes away with one hand it gives with the other. Although the political possibilities imaginable within the compass of the post-Fordist model of development tend to be limited, for example, to (1) whatever national-level class compromises can take place within the context of a global capitalist system and (2) local and

[1] The term "model of development" has been adopted from Lipietz (1992). In Lipietz's usage, the model of development incorporates three components: a regime of accumulation, a mode of regulation, and an industrial paradigm. Each of these components is defined in the discussion of Fordism below.

regional initiatives of capitalist industrialization and revitalization, these limited possibilities are *powerfully* suggested by the model, which has made it an inspiration for economic and social development initiatives at a number of scales.[2]

This necessarily means, however, that other possibilities remain unexplored. To take the example that most concerns me here, a politics of noncapitalist economic development is effectively precluded by the homogeneously capitalist nature of the post-Fordist imaginary. In the absence of a vision of economic and class difference, post-Fordist activists are installing a new form of capitalism on the economic and social terrain.

Fordism and post-Fordism[3]

The story of post-Fordism is often presented as part of a larger narrative of capitalist development, one that originated with several groups of French political economists known collectively to English-speakers as the "regulation school."[4] In this narrative, the industrial capitalist nations, and especially the United States, enjoyed a long period of prosperity in the first 25 years following World War II. This was the period of high

[2] Debates over progressive economic development strategies, new models of labor relations, and political directions on the left have been couched within the language of post-Fordism in a number of national settings, including France, where theories of regulation were first developed; the United States, where local economic development activism has recently burgeoned; South Africa, where planners are attempting to build a new economy for a new state; Canada, particularly in areas energized or dominated by the NDP; Great Britain, where journals such as *Capital and Class*, *New Left Review* and the now-defunct *Marxism Today* have been the sites of considerable debate over the economics and politics of post-Fordism; and Australia, where the existence of a Labor government throughout the 1980s and into the 1990s fostered a longstanding interest in post-Fordist approaches to reindustrialization.

[3] The ensuing narrative of Fordism and post-Fordism is not intended – nor could it possibly succeed – as an amalgam of all existing stories of Fordism and post-Fordism. It would be impossible to contain in one coherent narrative the various and conflicting contributions to these social representations. What I would say about my story is the following: Fordism and post-Fordism are certainly more than this, and they are often different from this, but they are also this.

[4] Michel Aglietta's *A Theory of Capitalist Regulation: the U.S. experience* is usually considered the founding text of Parisian regulation theory. The work of Alain Lipietz, some of which is cited in this chapter, provides a relatively accessible introduction to the Parisian school, though as Jessop (1990) notes, that school is increasingly difficult to view as a unity. For critical reviews of the complex and rapidly developing regulation literature that compare the different schools, see Boyer (1990), Dunford (1990), Jessop (1990) and Peck and Tickell (1994).

Fordism, in which mass production was coupled with mass consumption in a "virtuous circle" of growth.

The model of development espoused by the regulationists is centered upon a *regime of accumulation*, which is a stable allocation of the social product between investment and consumption. Under Fordism this regime involved continual innovation in the capital goods sector, leading to rapid growth in productivity and output that was balanced by an equally rapid growth in consumer demand. The relationship between production and consumption was stabilized or "regulated" by a *mode of regulation* that encompassed a wide variety of social norms and institutions. Large firms, mark-up pricing, wage-price indexing, and collective bargaining worked together to keep demand growing, productivity increasing, wages rising, and inflation creeping. New Deal-type social insurance measures like social security and unemployment compensation ensured that workers could continue consuming even when they were not working.

At home the unquestioned hegemony of the nuclear family (manifest in the notion of a "family wage") guaranteed a mass market with a voracious appetite for standardized consumer durables. And on the international front, Bretton Woods and GATT stabilized currency and commodity markets, making the world safe for capitalist investment and international trade.

The Fordist *industrial paradigm* is usually identified as mass production, undertaken in the context of a Taylorized labor process and a rigid division of labor for both workers and machines. In typical Fordist sectors like the automobile and household appliance industries, large batches of standardized goods are produced in a routinized, semi-automated production process using dedicated (or single-purpose) equipment and structured by the assembly line. Other familiar concomitants of mass production in its Fordist form are vertical integration, the pursuit of economies of scale, and just-in-case inventory control.

Fordism achieved its Golden Age in the 1950s and 1960s. By the early 1970s, it had met its social and technical limits and the long boom of the postwar period was coming to an end. Workers had become resistant to the stultifying efficiencies of the Taylorized labor process, which itself was resistant to further Taylorization. Despite the slowdown in productivity growth engendered by this dual resistance, consumption was unable or unwilling to keep up with the capabilities of the mass production system. The nature of demand was changing; niche markets for specialty and luxury goods were growing faster than mass markets for standardized products. Stagnating demand and import competition drove industrial capitalists to seek concessions from their workers. After years of collective bargaining and cost of living adjustments, they reverted to union-busting, take backs, and running away. In the US and UK, and later in other social formations, the state began to unravel the social safety net. Fordism had broken down.

In the wake of the crisis of Fordism the theoretical emergence of post-Fordism involved, among other things, the (contested) ascendancy of regulation theory in left social analysis and the uneven amalgamation of various other theoretical strands.[5] Perhaps the most generally accepted element of post-Fordism is the industrial paradigm, or model of industrial development, which comes not from regulation theory but from the literatures on flexible specialization and the Japanese style of industrial production.

Originally theorized by Piore and Sabel (1984), flexible specialization is now widely heralded as the solution to the technical and social problems of mass production. Technological change in the form of reprogrammable computerized equipment is revitalizing craft and small batch production, allowing producers to cope with the increased volatility and fragmentation of demand. Flexible workers and machines are replacing the dedicated equipment and deskilled workers associated with the old mass production system. In every phase of production, the just-in-time management system developed by the Japanese fosters upgrading, streamlining and cost control. Since workers are required to work in teams and to master a wide range of skills, rigid job descriptions and work rules – like the bureaucratic unions that devised them – are counterproductive and obsolescent.

In addition to reshaping the labor process and labor relations, flexibility has remade the firm and redrawn the map of industry (Storper and Scott 1989). Fordist mass production in an expanding economy fostered the growth of large integrated firms with a centralized decision-making structure and spatially dispersed production facilities. Flexibility, on the other hand, has promoted the emergence of smaller firms or independent divisions that interact with each other in a complex array of supplier and purchaser relations. The need to collaborate on design and production and to adjust quickly to changes in markets has given rise to new, more concentrated spatial arrangements. Small firms, for example, tend to cluster in "industrial districts" where they can share a pool of skilled workers, equipment, and services that none of them could support on their own. Major assembly plants tend to be situated at the center of a ring of just-in-time suppliers. Researchers point to the industrial districts

[5] See Amin (1994) for a collection of essays on post-Fordism that captures some of the diversity of approaches to its theorization.

of the Emilia-Romagna region in Italy – the famous "Third Italy" – and to the Japanese manufacturing system as existing examples of post-Fordist industrialization.

At the level of the macro-economy, and at the political and cultural levels, post-Fordism is less clearly specified. More pessimistic analysts (for example, Harvey 1989) characterize post-Fordist consumption as partitioned into luxury and mass markets by a polarized labor market: relatively few highly skilled workers enjoy "yuppie" affluence while a huge number of unskilled, casualized workers struggle at low wages to obtain the necessities of life. On the political level, a massive retreat of the welfare state coincides with the reassertion of market relations in all spheres of social existence. The culture of postmodernism elevates individual rights and self-development over collective organization and goals, replicating (and reinforcing) the decentralization and privatization taking place in the economy.

More optimistic leftists, including many of the authors in Hall and Jacques (1989a), see flexible specialization as the appropriate industrial paradigm for a highly productive society in which workers enjoy greater control in the workplace and communities can nurture and support a local (rooted and responsible) industrial base. They acknowledge that the politics of difference and democratic rights may be supplanting or transforming the politics of class; but they attribute this to the blurring of the line between management and workers in post-Fordist industry and to the rise of new social movements that are not defined by work-related concerns. In the cultural realm, they see postmodernism as a (sometimes complicit) challenge to commodification and to the restrictive elitism of high modern culture.

One distinctive optimist is Alain Lipietz, who criticizes both the pessimists and other optimists for capitulating to capital and ending the story before it has begun. According to Lipietz, the hourglass income distribution (described above) is an aspect of neo-Fordism, a retrograde attempt to salvage Fordism that has already reached its limits. He argues that the productivity increases derived from flexible production systems provide the opportunity for a new compromise between capital and labor, based on the latter's increased control over and involvement in flexible labor processes. Since the environment now poses strict limits to expansion, the new accord must involve worker demands for greater leisure and broader opportunities for non-industrial pursuits. These would improve the quality of life without requiring quantitative increases in consumption (Lipietz 1987b, 1988a and b, 1992).

The social totality in regulation theory: continuities and discontinuities in the Marxian tradition

According to theorists of capitalist regulation, models of development such as Fordism and post-Fordism coalesce as a matter of "historical accident." Regulation theory thus problematizes the processes of economic and social reproduction that other theories of capitalism have taken as their starting place or presumption:

> Starting from a certain analytic scepticism about systemic reproduction, the regulation approach asks how social relations take on stabilized forms – that is, how regulation occurs. Never assuming that reproduction of any social relation *must* occur, the task is to identify the institutionalized practices which stall contradictions and thereby contribute to reproduction. The definition of a system of regulation is that it regulates – i.e., stabilizes – social relations even though these relations are contradictory. (Jenson 1989: 72)

In the regulationists' understanding, then, a particular model of development is the product of a concrete historical situation, a unique concatenation of circumstances that allows for a sustained period of stability and growth.

In the aftermath of the crisis of Fordism, however, a visible inconsistency in this formulation emerged, as the activities of theoretical practice took on a contradictory relation to the prescriptions and pronouncements of the theorists themselves. As soon as Fordism was declared dead, theorists of regulation began the (theoretical) consolidation of the next model of development, putting the post-Fordist flesh upon the model's bare bones – the above-described regime of accumulation, mode of regulation, and industrial paradigm. In the process, the model of development was revealed to be an abstract social structure that exists outside its concrete manifestations. Independent of and prior to the actual social formations to which it gives form, the model of development is a social skeleton or template that organizes and arrays specific social existences and practices. Particular forms of capitalist society may be generated through "historical accident" and open-ended political and ideological struggles, but the general form of capitalist society is a theoretical given. This formulation locates the model of development within the strand of Marxist theory that could be called the "Marxism of the totality"[6] and

[6] On conceptions of social and economic totality that are prevalent within the Marxist tradition, see Cullenberg (1994b). See Hirst and Zeitlin (1991) for criticisms of totalizing social conceptions.

sets it against holistic conceptions of society as an articulated structure that is only partially closed and always under construction (see, for example, chapter 2 and Laclau and Mouffe 1985).

The structural essentialism that characterizes the theoretical use of the model of development and other similar social templates has had profound effects on Marxist conceptions of politics and of possibilities for social and economic transformation. The distinction between reform and revolution, including the denigration of the former and the millennialization of the latter; the association of revolutionary opportunity with moments of weakness or crisis in capitalist reproduction; the view of the working class as the privileged or natural origin of transformative socialist politics, and of mobilized collectivity as the authentic political form; all of these can be associated with the view of society as a stable and singular formation which is centered on a capitalist economy, principally fractured by class antagonism, and unified by a reflective reciprocity between economic, political and cultural domains. While these features of anticapitalist political thinking are both familiar and largely discredited, the social totality that attends and gives rise to them is alive and well, discouraging if not blocking the emergence of alternative visions and projections.

Centering

The principal center of the model of development is the problematic process of capital accumulation, which requires a mode of regulation (comprised of the full range of economic, political, and cultural norms and institutions) to stall its contradictions and permit sustained periods of stability and growth.[7] In the sense that each element of the complex mode of regulation is "focused" on the process of capital accumulation, that process is constituted as the social center.[8]

In many representations of Fordism and post-Fordism the model of development has a secondary or alternative center in the industrial paradigm. Mass production and flexible specialization operate not only as origins, but also as models, for other social practices and institutions; thus the Fordist image of standardization and the post-Fordist image of

[7] Indeed, the mode of regulation allows society to come into being in the sense of attaining the fullness, completeness and coherence that are associated with the Marxian and bourgeois concepts of development and that are enshrined (and revealed as myths of and desires for social unification) in modernist representations of a social totality.

[8] For an extended discussion of modernism in the Marxian tradition that examines the effects of centering (primarily with respect to subjectivity), see Amariglio and Ruccio (1994).

fragmentation/differentiation are translated from the realm of production to other social domains.

Ironically, the centricity of the model of development is most visible in moments of crisis and dissolution. Whether crisis originates in the contradictions of the accumulation process, or in the technical and social limits of the industrial paradigm (or whether it is transmitted to these central points from some other social location), it radiates out from these social centers to destabilize the entire economic and social formation. A crisis of the part thus engenders a crisis of the whole. In the process the crucial part reveals itself as the central social feature, around which the rest of society is arrayed and to which it is in some sense subordinated or subsumed. When mass production falters or intensive accumulation fails to proceed at a steady pace, the result is a crisis in the Fordist model of development, one that brings about a complete reconfiguration of the social space.[9]

Centered on capital accumulation, or upon the forces of capitalist production (of which the industrial paradigms of mass production and flexible specialization represent specific forms), society is a unified formation with a specifically capitalist identity. Thus the forms of the state, and of ideology and culture, are all specified with relation to capitalism and to whatever are conceived to be its defining features and problems (tendencies toward contradiction in the accumulation process, for example, or competition and profit-seeking at the enterprise level, or antagonisms between capital and labor).[10] By virtue of its central role in defining the specificities of the other social moments, "capitalism" becomes the dominant term in the social articulation. Crisis is therefore not only crisis of the totality but it is also *capitalist* crisis. This means, for example, that the (noncapitalist) economic relations in households could undergo transformation in a large number of domestic settings (such a possibility is not precluded by the model and might constitute a crisis in the sphere of the household) but these transformations would not be perceived as producing crisis in the social totality or as heralding the emergence of a new "model of development." On the other hand, when mass production gives way to flexible specialization in a large number of industrial sites, these local transformations are interpreted as the herald

9 In a decentered totality, by contrast, one might theorize a crisis in a particular social practice or institution (rather than a crisis originating in one site and spreading to the social whole) and trace the complex and contradictory relation of this crisis to changes in other aspects of society. An example of this mode of analysis is presented in chapter 9 where I trace some of the complex effects upon industrial production of a crisis in household relations.

10 These again are "necessary" features of an abstract capitalism that pose problems for concrete actors and institutions.

of a fundamental and thoroughgoing one (a "sea change," in the familiar Shakespearean allusion).

This has definite implications for the conceptions of politics and political subjectivity that are associated with the Fordist and post-Fordist models. On the most general level, the construction in theory of a coherent and unified social formation is associated with an implicit collective subject – a "we" whose unity is given by an abstract theoretical construct rather than (or in addition to) being constituted through political or cultural struggles. In the specific case of regulation theory, this collective subject is made up of "citizens" of a capitalist state who are "workers" in capitalist enterprises (or students being prepared for work, older persons retired from work, and household workers engaging in the reproduction of workers).[11] The very form of the regulationists' story – as an overarching narrative of capitalism in the post-war period – posits a fundamental commonality of experience as the limit of social difference.

Despite the regulationists' insistence that class position is no more consequential for political activism than other forms of social differentiation, the principal structuring social antagonism is delineated along class lines. Thus "national modes of regulation are characteristically treated as class compromises,"[12] implicitly privileging class actors over other kinds of actors (or class politics over other kinds of politics) in the conflictual constitution of social regulation.[13] Nonclass movements of social and

[11] For Piore and Sabel, the "worker" is also the implicitly privileged political subject. They suggest (1984: 303–7) that the transition from mass production to flexible specialization may be one of the preconditions of a new era of "yeoman democracy" characterized by free and independent producers and an egalitarian social structure.

[12] Hirst and Zeitlin (1991: 22). They argue that "at a methodological level, regulation theorists reject the idea of classes as collective subjects whose interests can be derived from the abstract structure of capitalist relations of production" but "when it comes to more specific regulationist analyses, social forces are analyzed in class terms with little attempt at empirical justification" (p. 22).

[13] Poststructuralist theorists of difference have pointed to the problematic implications of thinking society as a centered unity with a fixed and singular identity. For Laclau and Mouffe (1985), for example, "society" (in the sense of a fixed articulation of social practices, identities, and institutions) is an impossibility. In a social space that is multiply structured, multiply fractured, and that gives rise to conflicting narratives that compete for recognition and acceptance, new subject positions and identities are continually emerging; produced by and generative of new antagonisms and new political projects, they reconstitute and redefine the society that gave them birth. In the sense of providing closure around a central process or relation, or establishing fixed relationships among social elements, society is something that cannot exist; the identity of the social (as an objectivity and as the objective grounding of collective subjectivity) is figured instead as a negativity or an absence: "To understand social reality, then, is not to understand what society *is*, but *what prevents it from being*" (Laclau 1990: 44).

political identity (defined, for example, by gender, race, or sexuality) may mobilize politically around their own concerns but the politics of identity cannot alter society's fundamentally capitalist nature. Socially transformative politics is a politics of collectivity or coalition which is focused on stabilizing and ameliorating a capitalist economy or on constructing a socialist one (whatever that may mean) in the macrosocial space of the future.

The imperative at the center

In the literature on capitalist regulation, the "regime of accumulation" is unproblematically placed at the heart of a structural model of capitalist society. Capital accumulation is the central process of capitalist development and, as in many Marxian theories, its centrality to that development is asserted but not theorized.

What is this process of capital accumulation that has such unquestioned theoretical status? In its most narrow sense, capital accumulation refers to the growth of fixed constant capital and, more precisely, to the expansion of fixed capital value. More broadly it refers to the growth of productive capital, including fixed and circulating constant capital and variable capital. More commonly, it suggests the expansion and growth of capitalism, the incorporation of more branches of production, more people, and more space into a capitalist economy. (These senses may of course all contradict each other; so, for example, an increase in the value of fixed capital may be associated with a sharp reduction in the deployment of variable capital and a concomitant shrinkage in the capitalist workforce.)

Within many strains of Marxism it is traditional to confer a special ontological status upon capital accumulation. Of all the payments made out of surplus value by capitalist firms, that made to secure an increase in productive capital (and more particularly in fixed capital stock) has come to be seen as the most important and even as a necessary corollary of participation in a capitalist production process. Though capital accumulation is ultimately an activity of individual firms, it is more commonly construed as an inevitable concomitant of capitalist development, a "tendential law" of capitalist motion. When capitalist economies "break" the law of accumulation, entering prolonged periods of disaccumulation, the latter process is often seen as a contradictory expression of the accumulation process or as a necessary prelude to renewed accumulation – in other words, disaccumulation is seen as itself an effect (in the sense of requirement) of accumulation.[14]

[14] I am indebted to Norton (1986, 1988b) for these arguments.

The regulationists have rejected the ahistorical narrative of the immanent contradictions of capital accumulation, with its associated teleology of historical progress and succession. Yet capital accumulation is still at the center of their model of development. From that central location it not only contributes to a vision of social coherence but also (by virtue of the hierarchical ordering that is an effect of centering) takes priority over other social processes, becoming an imperative of the social totality rather than simply an activity of firms.

Though divorced from its association with an evolutionary narrative of capitalism's inevitable breakdown and supersession, accumulation brings its other meanings to the stories of Fordism and post-Fordism, which its status as a central process and systemic imperative cannot help but reinforce. Most prominently here I am thinking of the growth imperative that is traditionally associated with capitalist economies. If the regulationists have dispensed with the inevitability of capitalist breakdown, they have not dispensed with the inevitability of growth. Growth remains an unquestioned "law" of capitalist development, with the implication for progressive activists that politics must at least accommodate and at most foster capitalist expansion (the alternative to the "necessary" process of growth being a crisis of accumulation).

Like capital accumulation, the industrial paradigm functions not only as a unifying center of the social representation but as an integral element of a dynamic growth machine.[15] In the competitive struggle for survival and profits, capitalist firms are continually engaged in improving their technology and labor process in order to insure productivity growth, or in order to enhance market share by other competitive means; if their improvements are no longer aligned upon a single trajectory, and if competition no longer takes a single form, this does not mean that firms do not reflect growth tendencies and dynamics.[16] In different ways the various forms of post-Fordist theory install a capitalist growth imperative that cannot be contraverted. Local economic development activists attempt to capture the effects of this imperative within a particular community or region, while macrolevel political projects aimed at generating a new class compromise (for example, Mathews 1989b) situate themselves within the boundaries of profits and markets and their requirements for growth. In either case the economy is the site of an autonomous motive to which

[15] Even cultural theorists Hall and Jacques (1989b) participate in this technocentric vision, when for example they argue that "it is (flexible specialization), above all, that is orchestrating and driving on the evolution of this new world" (p. 12).

[16] While new theories of competition that do not reduce this process to price competition are characteristic of the post-Fordist literature, they do not explicitly dissociate the process of competition from the capitalist imperative of growth (see, for example, Best 1990).

the rest of society – including left politics – must accommodate and adjust.[17]

Correspondence and stability, contradiction versus noncontradiction

While the regulation school departs from the many forms of traditional accumulation theory that represent capital accumulation as the "driver" that drives itself (regulating through its own internal logic both itself and ultimately all other social processes and conditions), the centrality of accumulation to the model of development betrays the continued presence of an economic essence. The pull of the essence is attenuated, in the sense that the regime of accumulation does not call forth its corresponding mode of regulation, but its gravitational force is still detectable in the relationship of correspondence between the two components of the social structure (Ruccio 1991).

With the essentialism of this model, in which one part (the mode of regulation) comes to correspond with another (the regime of accumulation) and thereby becomes one of its phenomenal forms or expressions, comes the related "problem" of noncontradiction.[18] The mode of regulation functions to create stability in the relations between consumption and accumulation.[19] Through this theoretical construction, in which a presumed source of conflict and disruption is temporarily harmonized with its social context, many versions of regulation theory displace contradiction from the theoretical center stage. Following this path, theories of Fordism and post-Fordism tend to highlight the ways in which the economy and polity reflect and reinforce each other rather than the ways in which they contradict and undermine each other. For Harvey (1989), for example, flexible accumulation finds its cultural reflection in postmodernism and its political reflection in post-Marxist radicalism, neoconservatism, and the privatizing or entrepreneurial state.

Whether such reflections are presented as an overlay of cultural and political logics upon the logic of the economy (Jameson 1991; Harvey 1989), or simply as the refracted image of one social dimension at

[17] In this sense, rather than in the stronger sense of economic determinism, post-Fordist theory is economistic.

[18] Essentialism is here understood as the intellectual process of locating a single element (or a set of elements) at the center or origin of a process or event. With respect to causation, then, the essentialist move involves an attempt to isolate sole or principal causes (within Marxism, such causes are often economic); with respect to identity, essentialism involves the positing of a stable core such as gender or race (see chapter 2).

[19] Jenson (1989) refers to the mode of regulation as "stalling" contradictions; Aglietta (1979: 383) sees it as "mitigating" contradiction.

various other levels of social existence, the theoretical privileging of correspondence presents certain problems to potential political actors. Perhaps most prominently, it implies that political interventions that do not accord with the reigning model of development are unlikely to succeed. They will meet with resistance, not only in the specific social sites at which they are directed, but from all levels of society and culture. In this way reflection theories limit political visions and discourage (certain kinds of) political projects.

In addition to correspondence between the parts of the regulated/regulating whole, the regulation school emphasizes correspondence over time. History becomes a succession of epochs of stability separated by periods of structural crisis. Interestingly enough, they tend to do the same thing in space, suppressing complex contradictions in the "complete" social formations of the advanced capitalist countries only to see them arise in the "incomplete" periphery.[20]

The discursive emphasis on stability rather than change also has political implications. Theories of Fordism and post-Fordism reinforce the historical (and existing) tendency of leftists to identify moments of crisis as the appropriate times for political intervention rather than seeking out contradictions as a continual source of change. In creating a predictable development trajectory that counterposes long periods of capitalist strength to shorter periods of weakness and upheaval, the regulationists obscure the transformational opportunities afforded by alternative conceptions of contradictory and uneven development. By focusing on new forms of "class compromise" appropriate to the stabilization of a post-Fordist regime of accumulation, they become a condition of existence of post-Fordist regulation rather than the cornerstone of an anticapitalist political culture.[21]

The conception of periods of social coherence and stability punctuated by moments of dissolution and change is a familiar element of the left political tradition. Like the holistic vision of capitalism as a society-wide system, images of stability have long undermined leftists' abilities to engage in revolutionary politics, encouraging instead a politics of preparation or postponement. Whereas systemic holism makes it impossible to identify small or local transformations as revolutionary events, stability reinforces the idea that the energies of the left must be devoted to "reform" until the whole begins to crack, at which time the moment of crisis represents an opening for a true politics of economic and social transformation. In Europe, Australia, and the US many activists are now engaged in consolidating a progressive capitalism, in part because their theoretical framework constructs "fundamental" change as a distant

[20] See Ruccio (1991) for a discussion of incompleteness in the periphery.

rather than proximate possibility. Representing change differently might promote the emergence of very different political visions and projects. If contradiction and antagonism were seen as generating instability in every form and being, class transformation might be envisioned as a regular occurrence and noncapitalist development could become a focus of the politics of every day.

Representations of capitalism as political culture

Theories of post-Fordism – through their lineage in regulation and flexibility theory – have come to resemble the many forms of mainstream development theory that privilege industrial technology and capital investment as causes of capitalist growth. Like its regulationist forebears, the story of post-Fordism is "particularly attentive to the institutional fabric of a working growth ensemble" (Storper and Walker 1989: 203). It is this characteristic of post-Fordist theory as an economic development discourse that has made it attractive to certain segments of the left. Many progressive economic development projects are now inspired and informed by images of a post-Fordist society which incorporates the technical and allocational conditions for stable capitalist growth. As these discursive concerns become the focal concerns of political

[21] This is not to say that contradiction is absent from the discourse of Fordism and post-Fordism. One could say, in fact, that the discourse is structured around a set of contradictions, which take the form of simple oppositions. Thus a knowledge of the Fordist model of development allows one to predict quite precisely the contours of post-Fordism, not merely because the essential structure of the model of development is unchanged but because post-Fordism is defined both by the absence of Fordist institutions and practices and the presence of institutions and practices that are historically or logically "opposite" to those associated with Fordism. Perhaps the most familiar example would be the industrial paradigms of mass production and flexible specialization, which are generally defined by two sets of opposed characteristics in a formation that Hirst and Zeitlin (1991) identify as "ideal typical" but which – in the literature on Fordism and post-Fordism – are consigned to different historical periods and models of development. In this blending of historical narrative and categorical construction, history is represented as reflecting the structure of a conceptual opposition, and difference or contradiction is constructed via a set of exclusions (thus flexible specialization, for example, is logically specified in terms of its exclusion of the characteristics of mass production). Contradiction, then, is subsumed to a simple dualism ("difference between") rather than theorized in terms of "difference within" as a source of uniqueness, complexity, and continual transformation. What is occluded here is the possibility of change (of something into something else) signaled by dialectical conceptions that work against the stability of categories, indeed against certain of the logics of language itself. Such a dialectical conception can be found in Althusser's concept of overdetermination (see chapter 2).

projects and visions, the discourse of post-Fordism contributes to a "realistic" politics of progressive capitalist development. But by virtue of its theoretical emphases, it also contributes to the greater unrealizability of an alternative politics, one focused on noncapitalist projects of social and economic construction.

Perhaps the most obvious point of departure for an evaluation of the political effects of post-Fordist theory is the recent history of economic development activism. This activism gained momentum when the older industrial economies began to fall apart (or, at least, when their dissolution became visible in the 1970s), and left trade unionists and other activists formed militant coalitions to create and maintain jobs in communities devastated by disinvestment. In the US and elsewhere, they fought against plant closings, analyzed corporate disinvestment strategies, and formulated plans for industrial renewal. They campaigned for state aid to "mature" industries and for plant closing legislation at the state and federal levels. They engineered worker buyouts of closed plants. They put forward plans for rapid transit and infrastructure development, and other efforts to create employment, encourage investment, and improve community life. In the late 1970s and the 1980s, progressive organizations, the most famous of which is the Tri-State Conference on Steel, sprang up in all the older industrial regions of the US. Many of these organizations are now members of an umbrella organization called the Federation for Industrial Retention and Renewal (FIRR).

Among the activists involved in these projects, most of whom saw capitalism as the cause of deindustrialization,[22] visions of class transformation occasionally emerged. Proponents of "lemon socialism" advocated buying up capitalist discards and operating them as cooperatives. Staughton Lynd (1987) and others worked to establish worker and community rights to industrial property[23] and groups traveled to the Basque highlands to study Mondragon. Despite these occasional glimmers, however, most leftists struggled through the 1970s and early 1980s without a clear vision of the future.

It is into this visionary vacuum that the story of post-Fordism has

[22] In the sense that capitalism is defined by the imperative of profitability and an inherent tendency toward expansion beyond the borders of particular nation states. Thus, when the technological conditions of international communication and transportation had sufficiently developed, and competitive pressures increased from a variety of sources, capital became relatively footloose and abandoned the industrial heartlands of the Fordist era for greener pastures elsewhere (in one version of the deindustrialization story).

[23] See, for example, Singer (1988).

moved.[24] To activists long dismayed by the destruction of traditional industry, post-Fordism in general and flexible specialization in particular offered an inspiring model of industrial regeneration. Rather than a return to the prosperity of the past, this body of theory promised a new world in which small enterprises could thrive and workers could realize their human capacities instead of emulating machines. Suddenly, in the mid-1980s, those who had struggled to retain and renew large-scale industry saw a new road open before them. By promoting locally based post-Fordist development strategies, they could make their communities less vulnerable to multinational firms that milk and close profitable plants. By fostering "modernization" among small and medium-sized firms, they could increase their chances of survival in a competitive global economy. With the compelling image of flexible specialization as their model and guide, left opponents of deindustrialization became active proponents of economic development along post-Fordist lines.

Many left activists and government planners are now engaged in post-Fordist industrial development efforts, pursuing a number of paths. They may, for example, be promoting technological expertise among managers and workers, fostering the development of industrial districts, seeking out niche markets for backward firms, or managing training programs that emphasize worker involvement in labor process upgrading and technological change. They may also be involved in larger struggles to complete the transformation of the Fordist welfare state, or to establish the macro-economic conditions which would permit local economic development initiatives to succeed. As progressives, they understand these activities in a variety of ways. Some see themselves as creating the basis for a rising standard of living for all, while others are contributing a progressive voice to mainstream political activities, or preventing further job loss and union decline, or struggling for a new class compromise to take the place of the Fordist accord. Those more familiar with post-Fordist theory may see their activities as consolidating a progressive post-Fordism, which they may view as a precursor or necessary condition of a desired future socialism. In any case, under the (acknowledged or unacknowledged) influence of a powerful model of capitalist development, they are putting a new form of capitalism in place.

Post-Fordist theory, of course, is not the only force behind this agenda. Economic development activism, while it is inspired and guided by models of flexible industry, is also driven by the prevailing ideal of

[24] It was, in fact, Piore and Sabel's specific intent in *The Second Industrial Divide* to articulate a simple and attractive industrial paradigm that the industrial development community could mobilize around (Charles Sabel 1989, oral communication).

"competitiveness" that holds both the right and the left in its thrall. From the perspective of Robin Murray (1988), perhaps the pre-eminent post-Fordist development specialist in Britain, and his American counterpart, Michael Best (1990), we must be competitive in order to lay the groundwork for progressive economic alternatives. This requires the development of intensely productive, flexibly specialized, capitalist industries. Once again, the left (or part of it) has convinced itself that the only way to create a noncapitalist alternative is to create a prosperous capitalism first.[25]

It is not surprising that post-Fordist theory should encourage a progressive politics of capitalist development. Theories of post-Fordism offer a wealth of insight into the conditions promoting a stable and prosperous capitalism and very little insight into alternative forms of development, whatever these might be. The universality of capitalism within advanced industrial social formations is one of the grounding assumptions of post-Fordist theory, and of the activism that attempts to promote a progressive post-Fordism. Many activists are aware, of course – more than are the theorists whose work informs their activism – that they are constructing a new form of capitalism, but they are persuaded that there is no viable alternative. Not only is capitalism the only game in town but its socialization is the principal route to socialism (if that end is to be desired). To refuse to build capitalism is to build nothing at all.

Theories of post-Fordism, centered as they are on the conditions and consequences of the flexible industrial paradigm and stable capital accumulation, present a world in which capitalist development is the only road. With alternative entry points or narrative centers, different development avenues could perhaps open up, and different political subjects and projects might be called into being. For a moment, I want to consider the possibility of an alternative discourse of class, and explore some of its political implications.

Knowledge, class, and industrial society

Whether or not we wish to work within theories of post-Fordism, or other macronarratives of social and economic transition, it is interesting to consider the effects of taking class as the entry point to a knowledge of contemporary industrial society.[26] Such a knowledge might see social formations as complexly constituted, encompassing a wide variety of

[25] An important exception to this general picture is the London Industrial Strategy of the Greater London Council, which stands out as an attempt (though one thwarted by the Thatcher government) to acknowledge and build a variety of economic forms.

[26] The conception of class that is employed here is elaborated in chapter 3.

class and nonclass processes, and would permit an understanding of some of the myriad and contradictory ways in which class shapes, and is shaped by, other aspects of society. From this entry point our discourse might focus on the conditions and consequences of exploitation, rather than on the conditions and consequences of a particular industrial paradigm or regime of accumulation.[27] It could therefore illuminate a variety of points of intervention in processes of exploitation that are omitted or obscured in post-Fordist theory, and inspire political projects that are quite different from (though not antagonistic to) those that existing theories of post-Fordism have inspired. Let me suggest just a few examples of what this might mean. If these examples take a frustratingly abstract and sketchy form, that may attest to the difficulty of imagining a politics for which the language is currently underdeveloped and the social imaginary is not widely shared.

Distributive class politics

Fifteen years of public policy and progressive activism targetting particular industries and regions has focused on promoting healthy capitalist enterprises, whether locally based firms cooperating on the model of the industrial district or branch plants of multinationals lured by relatively cheap labor and a variety of subsidies and other inducements. Presumably these firms, which are supposed to create employment and generate regional income, are also generating a considerable surplus through the exploitation of labor. One might ask, to whom will that realized surplus labor flow?

It is likely of course that it will be parceled out among a wide variety of familiar destinations, including tax payments to the government, premium payments to insurance companies, management salaries and bonuses, shareholders' dividends, etc. Perhaps it will also be distributed to selected industrial workers in the form of bonuses and other incentives, while other workers are excluded from the prosperity it permits. Perhaps it will fuel the continued development of vast and fragile financial domains. Perhaps it will be reinvested in productive economic activity, but in some other region or country. On the other hand, it could also become the focus of what might be called "distributive class struggles" on the part of local citizens, workers, and government, efforts to capture some of the surplus for the region of its origin and the community it

[27] Exploitation in the strict sense entails the appropriation of surplus labor by someone other than the laborer; non-exploitative appropriation of surplus labor occurs when individuals or collectivities appropriate their own surplus labor to distribute it as they wish or require. Taken together, these processes of appropriation and distribution constitute the "processes of class."

supports. While such a possibility may seem outlandish at a time when firms are able apparently to demand almost anything of workers and communities, but communities and workers lack the reciprocal ability to make demands upon firms, it is important to remember that such general depictions of differential bargaining power always mask specific "exceptions," and that the depictions themselves rest upon a questionable view of power as *distributed* and advantage as *taken* (rather than, say, a view of power as *circulating* and advantage as always *up for grabs*). The practice of showering firms with subsidies that they are not required to repay, and of making one-sided bargains in which states or localities become obligated to private firms but those firms do not incur reciprocal obligations, are themselves constituents of the inequalities between firms and communities that activists, unionists and politicians bemoan (see chapter 8).

Recently in the United States state governments have become more concerned about reciprocity and more interested in penalizing firms for not meeting their part of the bargain (for example, by shutting down recently opened and heavily subsidized plants, or failing to create the promised number of jobs) (Greenfield and Graham 1996). This new attitude and practice suggests that firms are not entirely independent nor communities entirely dependent (in other words, that each may have leverage on the other) and lays the imaginative groundwork for a distributive class politics that is focused on obtaining allocations of economic surplus for local purposes, including (though not limited to) economic development projects of a noncapitalist nature. Subsidies and loans have sometimes been given to firms on the condition that they will reinvest a portion of their profits in the region and sector from which they were derived (this occurred, for example, in the case of the steel industry in the US, where a certain percentage of profits was required to be reinvested as a condition of receipt of government assistance). Certainly it might be possible (in either union or government negotiations with a firm) to arrange that distributions from the surplus be made to a fund that is allocated to local economic initiatives, the class nature of which is not specified but is potentially diverse. Thus, for example, a local development fund might be generated, supported by the company and matched by the union, but administered by the community, which could support a range of different businesses, employment possibilities, industries, and class processes. Projects could be solicited from the community by the administrators of the fund, who would be given a mandate to promote diversity not only in employment opportunities but also in the dimension of class.

In this way, for example, a community centered on a single industry like steel or coal could attempt to limit its economic dependence on the decisions of one firm, or the fate of one sector, and provide some conditions of economic mobility and diversity for its current and potential workforce (including currently employed, retired, and laid off workers in the principal industry, unemployed family members of those workers, as well as individuals totally unconnected to the principal source of employment.)[28] Such a community could increase the presence of noncapitalist economic activity and generate a discourse of the value of class diversity for economic sustainability. For just as one might want to widen the spectrum of potential industries and occupations available to community members, to increase the chances that the capabilities and inclinations of workers might find appropriate outlets in work, so one might want to increase the diversity of class processes and positions,

[28] In the coal-mining communities of Central Queensland, Australia, for example, which are classic single industry communities with a very narrow employment spectrum for women, and a very homogeneous occupational structure for men, the miners' union and the federal government have the opportunity, though not the vision that is requisite to the social possibility, to orient some of the surplus extracted from the miners to local economic development initiatives that might broaden and stabilize the regional economic base. Currently most of the union's efforts have been directed toward increasing the wages and benefits of the mostly male mining workforce (whose huge wage and benefit packages can be understood as incorporating a cut of the surplus, since they are so far above the compensation packages customary in other industries) and the federal government has both subsidized and taxed the industry (in the latter process, obtaining its cut of the surplus). These two types of distributive payments are the outcome of a class politics of surplus distribution (though one not called by that name) of a very ordinary and traditional sort. Currently this type of distributive class politics is associated with movements to stem wage growth or to cut business taxes in order to promote investment by firms, which amounts to a strategy (successful or unsuccessful) for promoting capitalist development by enhancing the pool of surplus value available for capitalist investment.

An alternative class politics of distribution in the coalfields of Central Queensland might focus on the allocation of grants, loans and non-financial forms of support to small businesses, including those that moonlighting coalminers are already involved in (such as hunting wild pigs and processing them as meat for the market) or that wives of coalminers are struggling to maintain (such as making children's clothing to be sold via mail order or at the outdoor tourist markets in coastal resort towns). These businesses are either self-run, with individuals appropriating surplus labor from themselves in what Gabriel (1989) following Marx has called the ancient class process, or they are collectively run and the workers jointly and communally appropriate the surplus. It is not unimaginable that the coal-mining companies could be induced to set up a fund to support local businesses, especially if the union was also to provide some of the funds. The mining companies in Central Queensland have a huge capital investment in these communities and therefore have an incentive to see these communities develop into viable places to live and work. At the same time, the companies themselves would not necessarily lose anything through these new distributive initiatives, which could be focused on redirecting the surplus that now goes into wage premiums for workers and tax payments to the state (see chapter 8).

to enable a richer set of options with respect to class (Gibson-Graham 1994a). (If it seems bizarre to talk about a "choice" among class positions, that attests to the aura of paucity and constraint that surrounds the discourse of class, and the prevailing sense that class is thrust upon us by a system outside which we have no existence and within which we have no purchase.)

Noncapitalist class politics

In addition to promoting distributive class struggles focused on obtaining cuts of surplus value from capitalist firms, left and community activists could think through more systematically the impact of changes and interventions in capitalist industries on noncapitalist class processes of surplus labor appropriation and distribution. This process of "thinking through" could unearth a wide array of possibilities for fostering class diversity. One obvious example, highlighted by post-Fordist theory, pertains to the increase in self-employment that has accompanied the process of rationalization and downsizing undergone by many restructuring Fordist firms. Often this self-employment is looked on by unions and other observers as a largely negative process of casualization: as employers outsource work that was previously done in-house, employees become involuntarily "self-employed." They lose their job security (becoming "temporary" and "contingent") as well as the benefits associated with full-time permanent status. Generally they get to work longer hours for less compensation, with no sick leave, paid vacation, or retirement provisions. Under these conditions, not surprisingly, many unionists have argued that the so-called "self-employed" worker is really a capitalist employee in disguise. It follows that unions, unable to take seriously the self-employed status of the casualized worker, have militated against outsourcing and casualization[29] and for the internalization of unbenefited and low wage "self-employed" workers within capitalist firms.[30]

An alternative perspective on this phenomenon has been articulated by Hotch (1994) in her work on the implications for the labor movement of the rise in self-employment. Hotch is interested in considering what unions (or other types of organizations not under the jurisdiction of the National Labor Relations Board) might be able to do for this group of people if they took them seriously as participants in an alternative, noncapitalist form of employment. Rather than seeing

[29] Not just in the interests of the worker, of course, but in the interests of the union, which loses membership as firms eliminate permanent employees.

the capitalist firm or government bureaucracy as the only providers of secure and well-compensated employment, labor organizations, she argues, could set themselves the goal of bringing increased security, compensation and opportunity to the growing sector of the labor force that is self-employed.

What this might mean has in part already been elaborated by strategists for the "new" or "nontraditional" workforce, who have argued that the labor movement should pursue a number of goals.[31] These include organizing "contingent" workers into bargaining units to press for higher wages and better working conditions; promoting universal health coverage and "portable" pension plans; and, most important though most difficult, attempting to get the rights of the self-employed recognized under the National Labor Relations Act (Hotch 1994: 60–1).[32]

Hotch brings to this strategic agenda the additional goal of promoting class diversity. Drawing on the model of the Self-Employed Women's Association (SEWA) of India, she distinguishes a variety of initiatives that might help the self-employed increase their "rate of self-appropriation," defined as the ratio of surplus to necessary labor. Increasing this ratio would augment the pool of surplus available for securing the conditions of self-employment (including payments to insurance companies, rent, investment in new equipment or training, and many other things) and might thereby make self-employment a more viable alternative to employment (and exploitation) within a capitalist firm. One way to increase the rate of self-appropriation would be to reduce necessary labor (or subsistence costs) through measures like provision of high quality day care at reasonable rates and good public transportation.[33] Another way would be to support the conditions of surplus labor production and appropriation by offering training programs (or access

30 This does not mean that there is no non-traditional approach to the contingent workforce, or an absence of progressive proposals for the worker who is not an employee of a capitalist firm, but rather that the dominant union discourse constructs self-employment as largely a negative rather than a positive option.

31 See, for example, Hecksher (1988).

32 Hotch (1994: 61–2) argues that the US labor movement should follow the example of the Self-Employed Women's Association in India (SEWA) which challenged the legal definition of a union in that country and won the right to be registered as a trade union.

33 These initiatives would not benefit all self-employed workers to the same degree (or at all) or in the same way, but the recognition that there is no essential commonality among such workers (except that they call themselves self-employed) and that therefore their interests are heterogeneous should not be allowed to stand in the way of positive initiatives. It must simply be recognized that such initiatives may provoke conflict within the union as well as enhance the well-being of some of its members (Hotch 1994).

to such programs) to upgrade the skills of the self-employed (who are often in highly competitive and rapidly changing fields like desktop publishing) or providing access to low interest financing for purchases of new equipment. Finally, labor organizations could help the self-employed establish purchasing cooperatives for raw materials and other inputs to the production process, or marketing cooperatives that would mitigate the difficulties of operating as a lone producer.[34] Hotch notes, for example, that

> in the United States, cooperative units could be particularly beneficial for workers providing business services to capitalist firms, many of which do not want to contract directly with self-employed workers because they are unable to provide "back up support" should they not be able to complete a project, or because they lack state-of-the-art equipment. By promoting cooperatives with greater purchasing power and the benefits of mutual support, labor unions can offer their members an alternative to subcontracting through a consulting firm, which appropriates a portion (or all) of the workers' surplus labor. (1994: 71)

What all these strategies for supporting self-employment attempt to do is to create the conditions under which individuals might appropriate their own surplus labor (rather than having it appropriated within capitalist firms) and at the same time enjoy a viable standard of living and decent working conditions. They also would promote noncapitalist commodity production and, more importantly, the existence of noncapitalist class processes as positive and desirable alternatives to capitalist employment and exploitation.

In addition to the case of self-employment, there are many other ways in which a post-Fordist knowledge could highlight noncapitalist class processes and their interactions with capitalist change. Consider, for example, the ways in which industrial restructuring interacts with the production of domestic goods and services in households, where surplus labor production and appropriation takes a noncapitalist form. The increased employment of women has destabilized entrenched patterns of exploitation in certain households, or at least made them into visible matters of concern. While some women have experienced employment in gender-segregated jobs as adding a new and exploitative class position to

[34] In an innovative experiment the Queen Vic Women's Centre (a non-profit women's organization) in Melbourne has organized a Women's Telephony Network to allow individual women and women's organizations to receive telecommunication services at a discounted rate. As a bulk user of telecommunication services women in this network can access not only discounted long distance telephone calls but also software packages for wordprocessing and small business accounting, thereby enhancing the organizational capabilities (both political and economic) of collectives and individuals (Srebrenka Kunek 1995, personal communication).

their existing exploitation as producers of domestic surplus labor, others have found that their experience of capitalist exploitation has given them leverage to create the conditions of communal surplus appropriation in the home, and still others have used their wage "slavery" to establish single-headed households where they produce surplus labor under conditions of self-appropriation (Fraad et al. 1994). An interesting and innovative approach to both industrial policy interventions and collective bargaining agreements might focus on the impacts of changes in industrial practices upon work and class relations in the home. In this light the practices of shiftwork and overtime potentially become very important, since the partner(s) of a shiftworker and overtime worker often have to increase their domestic surplus labor in order to maintain the economy of the household. By highlighting such household strains and inequities, a focus on household class relations could contribute to a new movement for a shortened work week, or flexible hours in the working week or day, thus reducing capitalist exploitation or pushing in the direction of such a reduction.

Finally, by virtue of their emphasis on teamwork, shared responsibility, and participation, post-Fordist forms of production may foster the conditions for the emergence of communal class processes in some industrial settings. It is important to consider the kinds of struggles that could promote the communal appropriation of surplus labor at the expense of capitalist development. How might the emphasis on financial incentives (for instance, bonuses, and profit-sharing) and on quasi-egalitarian worker participation models be "translated" from the discourse of competitive strategy to the discourse of class transformation?

None of these noncapitalist class initiatives is a remote or unlikely possibility, and in fact all of them are underway in various industrial and domestic settings, yet none is currently theorized as a concomitant of post-Fordist transition. This is in part because the narrative of post-Fordism is unambiguously a narrative of capitalist development, and one which is centered on manufacturing industry and capital accumulation rather than on exploitation and class. The post-Fordist narrative has been instrumental in the emergence of a new left politics of progressive capitalist development. By implication, a new politics of class transformation might be one of the effects of a new discourse and knowledge of class.

Conclusion

Underlying this analysis is a vision of knowledge as a fully effective constituent of social reality, and an associated vision of individual knowledges as necessarily limited and incomplete (in the sense of representing particular discursive and ontological commitments that must inevitably exclude other possible commitments). When knowledge is freed from the task of "corresponding" to an external reality (which it does not change and must not contradict) it enters a realm of heterogeneity and simultaneously leaves the realm in which some knowledges are adjudged more true or authentic than others because of their greater "affinities" with the real. Negotiating this world of difference, one may distinguish knowledges not by their degree of correspondence to the "real world" but by their singularities – among other things, the social and natural processes they problematize, the status they accord knowledge as a social process, the validation criteria they espouse, their disparate and contradictory social effects. From this perspective knowledges of Fordism and post-Fordism can be seen as political interventions: not only do they mobilize political subjects by constituting certain subject positions within the context of a social representation; they also establish the contours of a political imaginary by delineating a general social structure and a set of particular social forms, as well as the ways in which each of these can change. Theories of Fordism and post-Fordism thus contribute to a specific sort of politics, one that is given to certain hesitancies and urgencies, one that entails certain opportunities and foreclosures. It is in terms of these kinds of contributions, I have been suggesting, that these powerful social representations might profitably be assessed.

Most economic activism inspired by post-Fordist theory attempts to promote social equity through capitalist growth. This reflects in part the ways in which the centricity and correspondences of the model have contributed to a view of industrialized social formations as homogeneously rather than unevenly capitalist. If the economy is fundamentally capitalist, then successful economic activism must accommodate itself to that "reality" rather than pursuing the utopian chimera of noncapitalist invention.

At the level of the social template (that is, the model of development) society is a known entity, rather than one that is unknown and under construction. This positions political subjects as constituents of an objectively given social totality, one that can be modified (via a better or worse "class compromise") but that is ultimately something to which they must adjust. The limits of political subjectivity (both of its internal constitution and its possibilities of external expression or action) are established by an objective social structure – a capitalist society, animated by the imperative of economic growth, and constrained in its possible reconfigurations by an underlying structural essence that is not accessible to politics at all.

Alternative representations of society as a decentered, incoherent and complex totality could offer multiple points of intervention in class (and other) processes at any point in time. They could represent production as taking place in the household, the "informal sector," the industrial sector, the service sector, the state sector. None of these need be seen as the center of the economy or the locus of its principal driving force. All may participate in constituting the economy and the larger society; all harbor various technologies and organizations of production; all are the sites of class processes. And all are subject to change on a continual basis.

Such social representations might also have room for a different conception of class politics, divorcing the politics of class transformation from images of systemic transition. If class were dissociated from concepts of collective subjectivity and systemic development, a "personal" politics of class could potentially emerge, and the locus of class transformation might be translated from the theater of national industrial history to the individual household, firm, workshop, government office or to any other place where surplus labor is produced and appropriated. Flexibility, the principal byword of post-Fordist theory, could extend to the dimension of class rather than being suspended at the threshold of class transformation.

8

Toward a New Class Politics of Distribution

Class relations of exploitation have traditionally been the unquestioned target of a politics of class transformation, while issues of (re)distribution have more often been relegated to a politics of social democratic reform. This is a dualism that bears investigation as both forms of politics slide out of public view.[1] The privileging of exploitation over distribution as the truly legitimate focus of class politics reveals an essentialist vision of the economic totality as centered upon a core economic relation (between capital and labor) and a key flow of resources (the appropriation of surplus value) which, if changed, would revolutionize the whole.[2] In this vision, any intervention in relations not at this center may be socially just and worthwhile but could not fundamentally transform the economic system.

While leftist discourse has not entirely put to rest the reform/revolution dichotomy, mainstream economic commentators and policy makers have begun to read reform itself as a kind of revolution (and therefore as something to be avoided). In most industrialized nations the very effective social democratic manipulations of distributional flows in the economy are now under scrutiny for their presumed detrimental influences upon economic growth and survival. In contemporary popular representations,

1 Or are pushed out by an increasingly pervasive politics of pragmatism (also known as a politics of despair).

2 The classic class goal is the elimination of exploitation via, for example, the socialization of production. Whether such a change would not just reshape the ways in which surplus labor was produced, appropriated and distributed (by instating a different class process involving, for example, communal appropriation) rather than eliminate exploitation per se is a matter of theoretical and political speculation.

redistributions of social wealth are seen as threatening to undermine the foundations upon which economic security has been built. Their almost revolutionary capability of bringing into being a new society of welfare "cheats" and dependents is now repeatedly and urgently foregrounded in the attempt to salvage the system as "we" know it.

As a consequence we have seen, over the past twenty years, a concerted and growing attack upon a formal politics of (re)distribution. In the traditional sites where social wealth is condensed or collected, and from whence it is distributed or redistributed – the business enterprise and the state, for example, where struggles over the distribution of wealth have historically been legitimated and legally sanctioned – these familiar collection points are now seen as needing to be replenished rather than emptied, and the wealth they consolidate as needing to be husbanded rather than disbursed. In both private and public sectors, it seems, (re)distribution has become an unaffordable, unreasonable and inappropriate goal.

Distributional issues, with their ostensibly non-economic burden of social equity and justice, have been relegated to second place behind "real" and pressing concerns such as those of corporate profitability and national economic performance.

What is interesting to us is that both the positioning of a transformative class politics in opposition to a politics of social democratic reform, and of (re)distribution in opposition to economic growth, draw upon the same centered vision of the economic totality. And, what is more alarming, the discourse of economic and social centeredness that has contributed to the demise of an active politics of class is now threatening to extinguish the vibrant politics of social reform and redistribution that has flourished until now in many nations.

In this book thus far class transformation has been defined as the bringing into existence or strengthening of noncapitalist class processes of surplus *appropriation*. Now we wish to begin the process of imagining a class politics of *distribution* that explicitly addresses the possibilities of class transformation.[3] Rather than simply working for a reinvigorated but more equitable capitalism, distributive politics may conceivably contribute to a diverse economic landscape in which noncapitalist class processes are engendered by and coexist with capitalist class processes,

3 Our project in this chapter bears almost no resemblance to the specification of socialist distributional goals in Roemer's *A Future for Socialism* (1994). Roemer presents a blueprint for a market socialist economy governed by mechanisms that ensure distributive justice. Despite our different orientations, we recognize in Roemer's construction of a hypothetical socialist society that differs from those imagined or experienced thus far a shared desire to reinvigorate an interest in viable alternatives to capitalism.

and in which "unsustainable" economic developments give rise to sustainable growth as well as non-growth. Such a politics would give rise to, but also might be nurtured by, an economic discourse that is not so neatly structured by oppositions and exclusions (of production from consumption, capitalism from noncapitalism, sustainability from unsustainability, revolution from reform) and that draws upon economic theories and accounting frameworks that "decenter" the enterprise, the national economy, and other social sites.

Subordinated though they are, alternative theories and modes of accounting do currently exist – just as new forms of distributive politics exist and coexist with the old, which have certainly not disappeared. Drawing together theoretical developments, accounting taxonomies and political projects, we may model a discursive regime that enables a class politics of distribution different from the familiar ones that are currently under threat.

Distributional struggles

With the rise of industrial capitalism struggles over the distribution of social wealth became focused upon the wage relation – and specifically on the share of appropriated surplus value that flows as profit to the owners of capital or back to the worker as some sort of wage premium. Organized labor has traditionally fought to reduce the rate of exploitation of the workforce and to increase the wages and wage premiums paid to workers. This form of distributional politics has succeeded best where unions have been able to control the labor market for certain jobs or industries (often by using sexist, racist, and nationalist strategies of exclusion).

The decreasing power of organized labor in many industrial nations (measured in both numbers of union members and types and levels of militancy) has combined with dominant narratives of declining business performance to effect a retreat from a robust politics of distribution focused upon the profit/wage shares. Processes of restructuring have hit many of the industries and occupations in which organized labor has been most powerful and most able to exercise a legitimate claim on the distribution of socially produced wealth. The "labor aristocracy" as it was formerly constituted is now virtually a memory, and the distribution of income has dramatically altered, lengthening and trimming around the middle. Many labor unions can no longer bargain over wage increases without granting concessions in areas affecting the labor process or job organization, and in some sectors wage cuts or freezes have been agreed to by unions in order to maintain jobs. While a workplace-based politics led by class conscious unions was once seen as an important contribution to socialist transformation, today many emphasize the economism and divisiveness of past union battles (see, for example, DeMartino 1991).

These struggles consolidated the power and relative wealth of a core group of (largely male, white, primary segment) workers, and left unchallenged the primacy of capitalism. In any case the political goals of workplace activists have become considerably more modest. Maintaining hard won and barely adequate working conditions is a major preoccupation, along with recruiting membership from those now growing segments of the workforce (women, immigrants, young people) that were traditionally ignored by the union movement.

At the same time that there has been a retreat from a distributional politics focused on the wage/profit shares, there has also been an onslaught upon the state as a site of redistribution. Social democratic ideals of an economically just society, overseen by a strong state with the power to redistribute wealth, have been attacked as unworkable and undesirable by those concerned with economic growth. Not only are the mechanisms and results of an equitable distribution of wealth seen as unaffordable in the current context, they are seen as stunting the cultural orientation towards individual achievement – the necessary ingredient of economic success in the global marketplace. A politics focused upon distributional amelioration of the uneven impacts of capitalism no longer has credibility. Hence the dismantling in many industrialized economies of (the last) vestiges of the welfare state.

Both of these "retreats" from traditional sites of distribution and distributional struggles are shadowed by increasing concerns about social polarization. As the poor get visibly poorer – as shelters for the homeless are closed, school food programs are scaled back, and diseases of the poor such as tuberculosis re-emerge in advanced industrial countries – and as the rich get visibly richer, the impact of the declining middle class is beginning to impinge upon the daily lives of (at least) most urban dwellers. Yet while there is concern for poverty, for the exploitation of marginalized workers, for the abuse of economic power by financiers and other prominent business people, there are few public calls for a reinvigorated politics focused on wealth redistribution.

Admittedly this is a rather depressing and non-contradictory picture of the fate of an active politics of distribution, though it is perhaps for that reason the most familiar. It is possible, however, to paint a different and more encouraging picture, as recently new sorts of struggles over distributional issues have begun to take place. These new forms of distributional politics are linked by some common features: they draw upon the authority of alternative legal discourses to further the claims of certain groups to a legitimate share of social wealth; and they are focused on resource and property distribution rather than the distribution of income.

In the older industrial regions of the US, for example, the rights of stakeholders (including not only workers and managers but also suppliers, customers, service providers, taxpayers, and other community members) have been counterposed to the more narrowly defined rights of shareholders in communities faced with massive job losses due to plant closings (Singer 1988; Greenfield and Graham 1996), especially when a closing is seen as part of a corporate strategy to liquidate still viable productive capital. In some cases the recognition of stakeholders' rights by courts and public officials has contributed to innovative dispositions of industrial property that can be understood as distributions of social wealth from the owners of that property to the labor force and local community.[4]

A second example are the "aboriginal land rights" movements, which are an international phenomenon arising out of conflicts between indigenous peoples and others over the use of land and land-based resources such as minerals or animals, usually in remote areas. These movements have opened up the potential for distributions of wealth to traditional landowners in the form of compensation, rent and royalty payments on the basis of "indigenous rights" to the land.[5]

Finally, the movement for ecologically sustainable development has raised awareness of the need for a distribution of wealth obtained through nonrenewable or unsustainable economic activities toward renewable/sustainable ones and at the same time has ushered in another "new" form of distributional politics based on the "future rights" of succeeding generations to existing resources and environments.

[4] One of numerous examples is the bankruptcy decision in the case of Morse Cutting Tool in New Bedford, Massachusetts in which the judge awarded the plant to a lower bidder who promised to keep it open (rather than to the higher bidder who would close it), thus violating the owners' right to the highest price for their property while giving legal credence to stakeholders' rights.

[5] In Australia, for example, the passing of the Native Title Act in 1993 after the Mabo judgment made by the High Court has finally rescinded the longstanding legal claim of "terra nullius" by which the continent of Australia was deemed unoccupied when British colonization commenced in 1788. Aborigines who can establish a continued relationship to their land ("remnant rights") can now enter into negotiations to regain title and/or receive compensation for its use over the past 200 years (Bartlett 1993; Howitt 1994).

These three examples of new forms of distributional politics are quite different from the more established politics of distribution that focus upon the wage/profit shares and mechanisms of state redistribution. Though not couched within any specifically revolutionary rhetoric, they quietly offer the opportunity to contemplate different and even noncapitalist futures.[6] (This is not to say that the vision of a noncapitalist future has been entirely absent from the socialist-inspired union movement and from social democratic politics, but rather that such a vision has all but vanished from sight today.) What we have here is the irony that relatively marginal and quite local movements for stakeholders' rights, aboriginal land rights and sustainable development pose some of the interesting possibilities for a politics of distribution – one that could conceivably be linked to a politics of class transformation – while well developed and formally registered union, labor, and social democratic movements are complicit and acquiescent with respect both to capitalism and to redistribution.

It seems that the new movements are liberated to experiment with wealth distribution in ways that the more traditional movements generally are not. By contrast with an established politics that hinges on capitalist/labor relations and state redistribution, the new politics draws upon alternative discourses of rights and alternative visions of development, specifically challenging private property rights and the goal of economic growth. In abandoning the dominant discourse of property ownership and the essentialist tenet that growth is the necessary prerequisite to just distribution, these movements indirectly highlight the ways in which the more familiar distributional struggles have been confined within a particular discourse of economy. Within that discourse, distribution is a matter of equitable income flows (rather than stocks of wealth) and distributive equity is (increasingly) seen as depending upon

[6] Here we are not suggesting that different property relations would necessarily usher in noncapitalist forms of class relations (see chapter 3 which introduces class as a relation of exploitation involving the production and appropriation of surplus labor, and delineates capitalism as a specific form of class relation in which surplus labor is extracted from wage labor in value form). Class in our conception is overdetermined, rather than defined, by property ownership and other sorts of social relations. This means that property ownership might be one form of leverage that groups might use to promote noncapitalist class relations, but it would not by itself signal the existence of such relations.

We should also note that our vision of a noncapitalist future is not predicated on the general eradication of capitalism but simply involves the acknowledged coexistence of capitalist and noncapitalist economic forms. In other words, it is a vision of economic heterogeneity rather than of an alternative (noncapitalist) homogeneity.

meeting certain prior economic conditions.[7] Equitable distribution is thus positioned not only as a lower social priority but also as a dependent and derivative economic process. If the new distributional struggles have the potential to inspire a radical rethinking of distribution, it is in part because they are not situated within this reductive and hierarchical discursive frame.

The discursive positioning of distribution

Struggles around the wages share and state mechanisms of distribution have been conceived within an economic discourse that privileges a centered conception of the economic totality. One key flow of surplus value, that which is ultimately distributed toward increasing the stock of productive capital (that is, toward capital accumulation in Marxian, and investment in non-Marxian, terms) is privileged with greater influence over economic futures than others.[8] Profit generation is identified as the necessary condition of growth and prosperity. It is easy to see how this vision of the generative capacities of profit goes along with a conception of distributions to workers via wage premiums or to the state via corporate taxes as bleeding the economic system of its life blood.[9]

If economic health – understood here as entailing growth – is discursively harnessed to a particular distribution of surplus, then any distributional politics that diverts surplus to other destinations (whether

[7] Block (1990) makes this point forcefully and more generally: "Increasingly, public debate has come to hinge, not on what kind of society we are or want to be, but on what the needs of the economy are. Hence, a broad range of social policies are now debated almost entirely in terms of how they fit in with the imperatives of the market" (p. 3).

[8] We are referring here to a popular economic discourse prevalent in the business press and other media locations. It has its antecedents and affiliates within academic economics, including the Marxian strain. The discourse of profitability dominates many Marxian political economic representations of the capitalist economy where it becomes essentialized as a structural "logic." In particular it is drawn upon in the classical Marxian formulation of the crisis tendency within capitalism, the tendency of the rate of profit to fall.

The "logic" of profitability constitutes the core contradiction in the capitalist system (the source of its dynamism, its crises and restructurings), subordinating and discursively defusing the myriad other contradictions that might be seen to challenge capitalist reproduction (Cullenberg 1994b).

[9] In chapter 5 we discuss the power of organicist metaphors in representations of the economy.

within the national economy or the individual enterprise) is automatically situated in opposition to economic well-being. Essentializing one flow of surplus and endowing it with superior causal effectivity positions all attempts to divert economic surplus to other economic and social ends as stunting growth and foreshortening the potential for prosperity. Under a Keynesian discursive regime, with its more complex construction of the dynamic of economic expansion, distributions to wage earners (consumers of wage goods) and to the state were understood as contributing to both the growth and stabilization of demand. In the 1990s, however, redistribution in the direction of social equity is less likely to be seen as stimulating demand than as a drain upon potential investment, and redistribution to the wealthy is more likely to be condoned as a supply-side strategy to promote investment growth.

Of course, as recent genealogies of accounting practices show, the definition of what constitutes economically and socially productive and unproductive distributions changes over time. In the early twentieth century, for example, economic and social statistical monitoring procedures (such as standard cost accounting in business, grading and nutritional surveys in schools, and fitness and intelligence testing in the wider population) were devised within a dominant discourse of efficiency (Miller and O'Leary 1987). In Britain, the calculative technologies developed at that time ratified economic distributions to the state to improve the physical and moral fiber of the potential workforce, seeing this flow of social wealth as a necessary element in increasing the efficiency and productivity of the nation.[10] By contrast, within the dominant discourse of growth that prevailed in Britain during the 1960s, the state was encouraged to reduce distributions aimed at improving the muscle and productivity of the British workforce via individual and family-oriented policy, and to focus more upon direct subsidies to industry (such as depreciation allowances to reduce the tax burden on enterprises) in order to generate growth. The introduction of the "discounted cash flow method," a new economic-financial calculus to guide investment decisions of the firm, allowed all potential investments or distributions of capital to be scrutinized in terms of their net economic

[10] Within this social milieu the efficiency of business, the nation and the individual were assumed to be congruent. As Miller and O'Leary note, the "self-interest of the worker, employer and the social body alike, joined to the assurance of science, was to render the worker acquiescent in this 'taking hold' of his or her physiology, in order to experiment with it and to improve its productive capabilities" (1987: 261).

worth to the company (Miller 1991).[11] The complex effects of taxation and government incentives could now be factored into calculations of profitability and performance and a new vision of the "cost" to the individual enterprise of social redistributions could be gained. This new accounting practice allowed for the delegitimation of state distributions to social ends and the legitimation of government subsidies to business.

Changing discourses of national economic development and of the redistributive role of the state have differently constituted the rights and entitlements of economic subjects. Throughout much of the twentieth century, for example, women have been regarded as rightful claimants upon state resources in their capacity as unpaid domestic workers (wives) and breeders (Jenson 1986, Fraser 1989, Fraser and Gordon 1993).[12] Discourses of national efficiency combined with paternalism to ensure that women had access to a "family wage" or received state allowances or services that would ensure the health of the next generation of workers.[13] During the 1980s, however, under the aegis of the political champions of economic rationalism, the rights and entitlements of women were substantially reformulated. As Nancy Fraser argues, in the United States

[11] Britain's low rate of growth was attributed at this time not to an inadequate amount of investment, but to the quality of the investments made (p. 745). "The selection by firms of the appropriate investment opportunities came to be viewed as decisive" (Miller 1991: 735). "Not only investment in plant and machinery, but welfare and prestige investments such as gymnasiums, country clubs and palatial offices would be analysed by reference to the 'directional beam of capital productivity'" (Dean 1954: 121, quoted in Miller 1991: 742). "Departures from this beam were not necessarily wrong, but top management should be made aware of the *costs* of welfare or prestige projects. Such projects could be conceptualized as a cost, understood in terms of the amount of earnings foregone, just as easily as could other capital expenditures. The productivity of capital provided a principle which would allow realistic comparison of one investment proposal with another, summarizing in a single figure all the information relevant to the decision" (Dean 1954: 123, quoted in Miller 1991: 742).

[12] Jenson argues that the British state was concerned in the early part of the century that "interference in the form of income transfer would threaten the family, either by encouraging men to abandon their dependents and/or by hurting their pride" and that this was the rationale for distributing only services and advice (1986: 21).

[13] The notion of a "family wage" – a basic wage level that was calculated to be sufficient for one worker to support a wife and two children – was supported by many labor movements, particularly those dominated by unions representing male dominated primary labor segment workers. The "victory" of the family wage resonates with the paternalism of many distributional struggles around the wage/profit share. See Valenze (1995) for a recent review of the feminist literature on the origins of family wage discourse in the early 19th century, and Pujol (1992) for a discussion of the ways in which this discourse has been appropriated within economic theory.

pressure to be "'self-supporting' through wage work intensified" (1993: 12) and women who remained as recipients of state benefits or payments became stigmatized as "welfare mothers" implicitly positioned in opposition to "the reigning normative images of social order: work discipline, heterosexual nuclear family organization, female chastity, law abidingness, 'paying one's way' and 'paying one's taxes'" (p. 13). In Australia, since the recent implementation of policies recommended in the White Paper on Employment and Unemployment, all citizens are now to be members of an "active society" in which "'labour market participation is the key to full participation in society'" (Probert 1995: 106). For the first time in the nation's social democratic history, spouses are only to be constituted as legitimate claimants upon state resources if they are "job ready" (Probert 1995: 106–7).[14] The new "national common sense" (Fraser 1993: 8) prevailing in many countries identifies distributions via the state to women as reproducers as no longer socially or economically "productive" but as something to be curtailed at all cost.

Genealogies of accounting practices and social policy not only illustrate the subordinate positioning of equitable redistribution within a discursive regime that privileges investment or accumulation, but also suggest the extent to which this positioning has become a powerful orthodoxy influencing policy and politics in recent years. Today the belief that the distributive potential of an economy must be dependent upon or subordinated to its potential for growth – and thus that a society can only sustain an equitable social distribution of wealth if its economy is healthy and growing – is virtually unchallenged.

This belief links the apparent rejection of any distributional imperative within public economic life to the rise of a popular affective discourse of profitability and productivity and the retreat from a worker-focused distributional politics in the capitalist workplace. At present, there appears to be an unquestioned sense, both intellectual and emotional, that the generation of profit is a necessary and legitimate process. So when profitability is threatened from whatever direction (whether via international competition, fluctuating exchange or interest rates, rising wage bills or industrial disruption), "we" are all implicated and affected.[15]

[14] While women are finally to be constituted in their own right as economic subjects separate from their spouses (a cause for feminist celebration), the privileging of their willingness to undertake paid work in the constitution of economic citizenship is cause for concern.

[15] There is something socially compelling about the index of profitability – something that is absent when one mentions rates of capacity utilization and that is different to the emotions stirred by talk of exploitation.

Public economic discourse increasingly invades personal and private spaces, hailing "us" as either causes or co-victims of economic stagnation and instability. It is via "our" affective relationship to profitability, socially constructed as a legitimate individual/corporate expectation, that we become implicated in "what is to be done?" about the economic state of the nation.[16]

The urgency of debate about profitability and national economic performance has been exacerbated by recent popular representations of the capitalist industrial enterprise as a victim, positioned at the mercy of the global financial sector.[17] Even the multinational corporation, that paragon of all that is big, powerful, worldly and self-important, has suffered an identity crisis in the face of fluctuations in currency markets and the vagaries of share transactions. As victim, the MNC is prey to those institutions which, while not themselves owning or controlling productive resources, command paper assets and thereby determine who will, at least momentarily, "own" corporations and benefit from distributions of investment funds.[18] Corporations and their workforces

[16] The unquestioned status of profitability as an appropriate public indicator of economic performance has developed along with a "vast series of regulations and tools for the administration of entire populations and the minutiae of people's lives," the procedures that Foucault designates "disciplinary power" (Miller and O'Leary 1987: 238). That monitoring profitability has become a socially accepted aspect of economic self-management, just as monitoring school grades and fat intake have become aspects of individual self-management, is the result of a complex history of mediation between the "private" world of business and the "public" gaze of bureaucrats and regulators.

[17] This popular representation of the corporation as victim is mirrored in recent writings by economic sociologists. Stearns and Mizruchi argue, for example, that various theorists today, including finance-capital, organizational, resource-dependence, transactions-cost and agency theorists, see the corporation as dependent on external financing – operating, that is, with very circumscribed autonomy. This is in contrast to the once dominant managerialist view that "large corporations had become powerful, independent institutions controlled by inside managers, who were free from the constraints of stockholders and financial institutions" (1993: 279).

[18] The MNC's once unparalleled power endowed by ownership and control of the means of production, and consolidated by the global mobility of investments, appears to have been diffused or transferred to another place within economic space (see chapter 6). And it is unclear whether the financial institutions within this economic space even have the interests of "capitalism" at heart. As Bluestone and Harrison note: "Now detached from its necessity to abide by the laws of capital accumulation the finance sector oversees and facilitates capital movement into speculation, mergers, acquisitions – 'financial gamesmanship'" (1988: 54). The threat to existence may even reside within the MNC – share trading can dissolve corporations over night, productive investments can be milked for cash to play the currency markets and managers of corporations can be asked to trade themselves out of a job (Coffee, Lowenstein, and Rose-Ackerman 1988).

are positioned in this representation as somehow at fault for their low productivity or declining levels of profitability, and governments (at both the provincial and national levels) are targeted for public scorn over economic mismanagement.

As the public obsession with profitability and performance has expanded, so the representation of enterprise profitability has been simplified – streamlined, almost – into a vague but powerful notion of "what companies need in order to be attractive to investors." To service this new "politics of corporate need" belts are tightened, public services are privatized, social expenditure is cut and all manner of distributional mechanisms are legitimately, it seems, "downsized."[19] These actions are supported by a vision of the enterprise as an unproblematic reflection of the economy-as-a-whole, similarly ordered and centered upon a single imperative (the generation of profits and thereby growth). In this vision enterprises share a common structure and behave as the "universal calculating subject" (Cutler et al. 1978: 129), expressing at the individual and concrete level the abstract rationality said to be characteristic of capital as a system and/or class.[20] Such a vision precludes any consideration of the possible multiplicity and diversity of enterprise structures and the specificitics and irrationalities of corporate subjectivity.

To reinvigorate politics around distributional issues it may be necessary to denaturalize the economic discourse that situates distributions of social wealth in opposition to economic survival. It may be necessary, that is, to think within a radically different accounting regime that does not draw upon a centered vision of economic totalities, an essentialist understanding of economic dynamics and a conflation of the identity

[19] In the face of this dominant economic discourse about profitability and the detrimental impacts of social distribution upon national survival/development, traditional distributional struggles are not the only ones that are curtailed. The alternative distributive movements discussed at the outset of this chapter may also come under threat as they appear to make further costly demands upon an already depleted economic system. While perhaps desirable, such things as stakeholders' rights, indigenous peoples' land rights and the rights of future generations can easily be dismissed as simply unaffordable in the current conjuncture.

[20] Cutler et al. describe the conflation that takes place in this reduction as follows: "[In] . . . both Marxist theory and marginalist/neo-classical theories . . . the enterprise or economic agent is a universal calculating subject (by such a subject we mean an entity whose attributes and actions are identical to members of the class of beings in question). For a universal subject of calculation to exist a domain appropriate to that calculation must exist, a domain which is homogeneous and general (which mirrors the identity of the subjects and offers no obstacles to it). This means that in order for all enterprises or agents to use the same given calculative criteria in the same way they must all be of the same organisational form and encounter similar conditions of operation" (1978: 129).

of all enterprises with a singular structure and subjectivity, that of the universal rational calculating subject.

Abandoning the centered enterprise and economy

Fortunately, the task of rethinking the economic entity and its dynamics, especially at the level of the enterprise, is already underway. Coexisting with an essentialist discourse of profitability that constitutes a singular and centered enterprise is an ever-expanding literature generated within political economy, economic anthropology, the "new industrial geography," the "new economic sociology," and the "new institutional economics," which documents and debates the very different forms of industrial organization and corporate behavior that have emerged as part of contemporary enterprise culture.[21] Given the explosion of diversity in enterprise structure and modes of performance that this literature describes, it is almost impossible to talk of the "capitalist firm" as something self-identical, since there no longer appears to be – if there ever was – any organizational form, management culture or competitive position that can be identified as typical, ideal, dominant, or more efficient. Corporations are now seen to be "embedded" in a range of social, cultural, and local relations that create quite unique pressures and rationales (Granovetter 1985; Clark 1994; Mitchell 1995). Salais and Storper observe that

> there are multiple coherent economic logics of carrying out production . . . This combination of multiple, overlapping and, sometimes, conflicting determinants of economic logic opens up a non-trivial role for diversity itself. (1992: 189)[22]

Organizational and behavioral economists offer another counter-representation to that of the enterprise centered by one crucial flow or dynamic. Against the highly abstracted and singular theory of the firm, Cyert and March pose

[21] This literature differentiates capitalist enterprises on the basis of size, forms of ownership (public, private, sole proprietorship, or joint stock), geographic scale of operation and ownership (regional, national, or international), level of sectoral concentration or conglomeration, market power (competitive, monopoly, oligopoly), management culture and style, place in the product cycle, organizational design, socio-cultural embeddedness, economies of convention, and so on.

[22] Having prepared the way for corporate diversity, however, Salais and Storper attempt to systematize it into four worlds of production "distinguished by fundamentally different organising and operating principles, rather than different choices based on the same optimising principles" (p. 171). Corporations thus occupy either the Marshallian Market World, the Network Market World, the World of Innovation or the Industrial World.

> a perspective that sees firms as coalitions of multiple, conflicting interests using standard rules and procedures to operate under conditions of bounded rationality . . . (1992: xi–xii)

They dispatch the dominant discourse of profit maximization with enviable economy and efficiency:

> Perhaps the simplest attack on profits as a motive is also the most destructive. We can argue that entrepreneurs, like anyone else, have a host of personal motives. Profit is one, perhaps, but they are also interested in sex, food, and saving souls. It is rather difficult to deny the proposition, but if we accept it as critical, it is not easy to see how to devise a theory of the firm in anything approximating its present form (or even with its present goals). (Cyert and March 1992: 9)

Spurred by their interest in the enterprise as a site of many different behaviors and motivations, these economists question the "truth value" of fundamental economic concepts and accounting conventions:

> To what extent is it arbitrary, in conventional accounting, that we call wage payments "costs" and dividend payments "profits" rather than the other way around? . . . It makes only slightly more sense to say that the goal of business organisation is to maximize profit than to say that its goal is to maximize the salary of Sam Smith [an individual employee]. (Cyert and March 1963: 26, quoted in Neimark and Tinker 1986: 375)

Recent work in organizational theory and studies of management culture emphasize the various conflicting models of corporate leadership and strategy that vie for dominance within the firm. The ultimate path of corporate action can be seen as the complex result of many interacting and contradictory tendencies within an organization, and not as the working out of a dominant imperative or dynamic.

These various non-Marxist projects of complicating the view of the firm and its behavior are mirrored in the work of certain anti-essentialist Marxist theorists engaged in an epistemological critique of the role of centered totalities in Marxian economics (see Thompson 1986; Cutler et al. 1978; Resnick and Wolff 1987; Daly 1991; Amariglio and Ruccio 1995a). Thompson advocates a vision of the firm as a fragmented and decentered site rather than a presumptive unity:

> Instead of conceiving of the enterprise or firm as a relatively homogeneous, organic, functioning *unity* typified by a universal calculating subject ("management"), the suggestion is to conceive of it as a heterogeneous non-unitary, dispersed and fractured entity or social agency. This way of conceiving of the firm would then analyse these institutional forms as the "site" or locus of a combination of social mechanisms and calculating practices which are juxtaposed and articulated at that "site" but where this combination is not analysed as a unity. (1986: 176–7)

In similar fashion, Resnick and Wolff resist the prevalent tendency to understand the firm as a unity governed by a dominant principle or centered upon a fundamental process:

> Conceived as an overdetermined site, no one of the enterprise's economic processes (say, the economic process of accumulating productive capital in the example of a capitalist industrial enterprise) is more important, more essential than any other process in governing its development. (1987: 168)

An anti-essentialist discourse of the enterprise and profitability yields a very different knowledge of profit as a discursive artifact produced within different regimes of accounting:

> For some firms (profitability) is an objective which is closely controlled while for others it is more of a residual item when all else has been accounted for. How can there be an essential and unambiguous objective of maximization of profit for all firms if there is no clear agreement on what profit actually is? (Thompson 1986: 181)

> Even such categories as "profit" cannot be accepted as unproblematic in our understanding of the different aspects and movements of complex enterprises. Corporations make a whole range of investment, production and organizational decisions which will respond to many different kinds of considerations – political stability, labour opposition, social protest, and so on – beyond the simple calculus of profit in the development of wider strategic aims. The economic space, therefore, cannot be understood in terms of the unfolding of a single logic capable of unifying all identity. (Daly 1991: 88)

> If universal criteria of calculation are to be posited it is necessary to have a universal standard of measure of returns to enterprise. [But] it is, of course, well-known that concepts of returns to enterprise, such as profit, and of rates of return, such as profit-rates, are subject to a plurality of standards of measure. For example, the recent [Sandilands] report of the Inflation Accounting Committee . . . referred to a number of different concepts governing the measurement of the value of non-monetary assets, of profit, of stock, and of capital maintenance. (Cutler et al. 1978: 133)

Not only do different accounting regimes generate different conceptions of key terms, suggesting the contingency and ultimate arbitrariness of indicators of economic performance, but each of them constitutes individuals as economic subjects through a process of interpellation:

> Accounting information itself is meaningless, it can only have meaning if individuals choose to allow it to have meaning for them; if they submit to the discipline of the accounting system and make themselves accountable to it. This is the source from which accounting derives its real power, not

from the law courts, the stock exchange, or from the vigilance of elected officials. (Kelly and Pratt 1992: 20)

Corporate subjects are overdetermined by the discourse of the centered enterprise that grounds the truth and "accuracy" of the prevailing accounting system, allowing it to dominate corporate activity. Against this discourse of rationality and unified subjectivity, alternative accounting systems and different, possibly decentered subjectivities have little purchase upon corporate decision-making and action. Yet it is clear that the diversity and contradictory nature of enterprise structures and accounting practices invite us to speculate about the suppressed or marginalized subjectivities that reside within the corporation.

These disparate examples of economic radicalism tempt us to explore the implications of economic decenteredness for rethinking distribution. Both on the enterprise level and at the level of the economy as a whole, a decentered and disunified vision of the economic entity could free distribution from its traditional position of subordination to exploitation and investment/accumulation. Such a vision has the potential to liberate multiple economic subjectivities now trapped within the circumscribed domain of the universal calculating subject. Our question becomes, how might the abandonment of a centered economic totality and of an essentialist conception of economic dynamics and subjectivity allow for a less constrained role for distributional struggles?

Repositioning distribution

Within the alternative anti-essentialist accounting framework developed by Resnick and Wolff (1987) the capitalist industrial enterprise is discursively constituted as a site at which surplus value is produced and momentarily collected before being distributed in a multitude of directions. As such, this site can no more readily be seen as a place of solidity/consolidation than as a place of emptiness/evacuation. The enterprise operates as a conduit through which flows of wealth (including, amongst others, flows of surplus value) are continually generated, collected and dissipated. No one type of flow can be seen as always subordinated to other flows, rather all flows exist in relations of overdetermination.

If we think of the enterprise as an entity defined as much by the dispersion of social labor as by its condensation, it can be likened to the body in which identity has been freed from the defining and governing function of the "mind." Without mentality in a position of dominance over other bodily processes, the body and its identity may be seen as equally determined by all corporeal processes in their multiplicity:

> Selves become *dissipative systems*. It is not that all identity disappears on this model; but rather that identity has to be understood not in terms of an inner mind or self *controlling* a body, but in terms of patterns of potentialities and flow. (Battersby 1993: 36)

Thinking of the enterprise as a pattern of "potentialities and flow" opens up a complex and prolific world that could be analyzed in myriad ways. The entry point that Resnick and Wolff adopt, and that we adopt here because of our interest in reinvigorating a politics of distribution, is that of class.[23] They distinguish the flows percolating through the capitalist enterprise in class terms as (1) nonclass flows of income and expenditure and (2) flows of appropriated surplus value that are distributed toward (a) the purchase of more machinery, raw materials and labor so that production can be expanded and/or to (b) securing the economic and noneconomic conditions of existence of capitalist exploitation. For Resnick and Wolff the enterprise is an overdetermined site where economic and noneconomic, class and nonclass processes operating both within and without the enterprise constitute and reconfigure each other in contradictory ways. They do not, for example,

> understand distributions of shares of surplus value in the form of salaries to advertising managers, rents to landlords, taxes to the state, dividends to stockholders, and interest to bankers to be necessarily a "drag" on or a "barrier" to an industrial enterprise's growth and development. Instead, [they] understand such distributions in general to be conditions for its continued existence and development. (1987: 319)[24]

In their view there is no necessity for particular distributive flows either within or out of the enterprise to detract from its successful operation, nor is there a necessity for the enterprise to appropriate surplus labor or to accumulate in order to survive. Indeed, Resnick and Wolff's framework makes it possible to trace the way in which distributions of surplus value *away* from accumulation may enhance business returns. A company may, for instance, pursue a strategy of disinvestment in

23 Obviously there are many other entry points that could be adopted as ways of analyzing the enterprise as a decentered site, such as power, culture, risk, knowledge, race. The list is infinite.

24 The authors are clear that they are making a theoretical rather than an ontological commitment here – they wish to emphasize the positive and developmental effects of surplus distribution rather than the potentially detrimental effects of this contradictory process.

production that results in disaccumulation and reduced appropriation of surplus value, but at the same time receive nonclass-based revenue flows from property speculation or money market transactions that counterbalance the loss of class-based revenue and maintain the enterprise in the black. Alternatively a firm may run at a permanent loss in terms of class-based revenues and survive by drawing upon cultural or family loyalties and commitments. Producing a discourse of the indeterminacy of enterprise transactions and power relations is one step toward loosening the constraints placed by conventional economic discourse upon distributional struggles, one which makes easier the task of imagining alternative economic subjectivities and "rights" to distributed shares of social wealth.

Resnick and Wolff's decentered conception opens up many different visions and points of intervention around issues of class. Capitalist exploitation (the appropriation of surplus labor in value form) recedes as the privileged focus of class politics. As the moment of condensation of surplus labor, exploitation can also be seen as enabling a moment of dispersion and distribution.[25] The firm can be envisioned not (only) as a point of retention/consolidation but as a point of dispersal/evacuation, the institutional site of distributive class processes in which appropriated surplus value is disbursed to an openended list of social destinations. Many individuals and institutions – advertisers, managers, labor unions, local and national governments, banks, financiers, local communities, charities, accounting firms, etc. – make claims upon the surplus value and other funds distributed by capitalist enterprises. A vision of the firm as the decentered site of emptying and dispersal suggests the realistic possibility of an enterprise-focused politics of distribution, one that engenders and expresses new sorts of rights and claims (say, of workers and community groups) to distributed shares of appropriated surplus value. But the academic, popular and political discourses of what we have called economic centeredness constitute a major barrier to the creation of such an enterprise-oriented language and politics of distribution.[26] In the remainder of this chapter we critically analyze the subject positions and politics that emerged within an enterprise-focused regional struggle in Australia, attempting to sketch the limits, as we see them, placed upon a left political imaginary by the discourse of economic centeredness.

25 See chapter 3 for a conception of class processes involving two separate moments: an exploitative moment in which surplus labor is appropriated, and a distributive moment in which it is paid out to various social destinations.

26 See Amariglio and Ruccio (1994, 1995a) for extended discussions of the role of centeredness and its association with principles of order and certainty in modernist Marxist and non-Marxist economic and political thought.

Representations and regional futures

Corporate distributional strategies are closely and publicly examined and challenged when they have major regional implications. In the context of plant closures and regional rationalization of operations many corporations have been forced to make their corporate strategies and accounts more public and have been quick to represent themselves as organizations with coherent and rational plans for their own economic survival. To justify closure or massive layoffs the corporation commonly represents itself as a victim of high wage bills, low rates of productivity, high interest rates and shrinking international markets. In a number of cases, as we noted earlier, this familiar corporate self-representation has been challenged. Communities and workforces have drawn upon a discourse of stakeholders' (as opposed to shareholders') rights to force companies to continue production or to sell or divest their industrial plants to the local stakeholders. In many more cases, however, the local community and politicians are taken in by the corporate representation and agree to the survival measures it dictates.

Recently, the discourse of stakeholders has been commandeered and reconfigured by corporations and local economic development agencies interested in implementing a post-Fordist model of industrial growth (see chapter 7). The identification of stakeholders in local industrial development is an attempt to draw local communities and workforces into a "partnership" with business and the local state to ensure that corporate growth (and presumably, local employment) is secured. From the perspective of the corporation as a decentered and overdetermined totality, the dangers of this strategy for local stakeholders are many. Concessions granted by a local workforce and community as part of a "partnership" may not necessarily result in increased or continued employment, or contribute to the general economic well-being of the locality, but may provide the opportunity for a range of distributions with no particular regional orientation.[27]

The recent fortunes of BHP Ltd serve as a good illustration of the

[27] For as Thompson argues, "there is no necessary relation between funds that are available for investment in productive employment creating activity within the firm and the actual investment *undertaken* in such activity by firms. Firms have been net lenders to other sectors and this is because they have calculated in a particular manner (admittedly under rather specific conditions). They have certainly not been concerned with the reproduction of their labor power as a whole (despite some marginal but overemphasized hoarding of important kinds of labor which might be in short supply). Nor will the simple continuation or supplementation of such subsidization necessarily mean that they do become so concerned. They may well indirectly 'lend-on' any such increased subsidization particularly if it takes the form of supplementary capital allowances etc." (1986: 120).

manipulation of discourses around distributional corporate and state strategies. To any Australian, BHP represents the quintessential capitalist corporation. Whether associated admiringly with gilt-edged shares, huge profits and a decisive role in national and regional development, or critically with paternalism, adversarial industrial relations and regional insecurity, this business entity is a major presence in the national (and increasingly the global) economy whose identity appears solid and above all powerful.[28] The basic nature of its industrial activities – iron ore and coal-mining, off-shore oil and gas extraction, steel making, metal fabrication – conveys a sense of permanence in and importance to the productive fabric of the national economy. The magnitude of its reported profits and its command over public representations of its corporate strategies via advertising and the business press contributes to a picture of an organization unambiguously centered upon accumulation and growth. And the course of recent regional history in Newcastle and Port Kembla-Wollongong (both sites of BHP-owned steel plants) as it has been reported to the public and represented in corporate accounts contributes to the vision of a corporation engaged in an orderly and strategically commanded process of restructuring to enhance its international competitiveness (O'Neill 1994: 180, 204).

During the 1980s and early 1990s BHP was a site of struggle, strategy and power differentials. Pervasive and extremely powerful representations and self-representations of BHP as a rational calculating subject were used to enable the corporation to set the terms of large-scale state redistributions of wealth in its favor and to skew state-condoned renegotiations of labor relations to its benefit (Fagan 1987; O'Neill 1994). In the two oldest steel-producing regions the results have been devastating in terms of job loss and regional instability.[29] Because the corporation's activities have had major impacts on regional economies in

[28] Broken Hill Proprietary Limited (BHP Ltd) is Australia's largest industrial business organization. It began in 1885 as a silver, lead, and zinc producing and exporting company at the famous Broken Hill mine. Using profits generated from this rich lode the company switched to the production of iron and steel in 1915. By 1935 it had become the monopoly steel producer in the country with steel mills on the New South Wales coast at Newcastle and Port Kembla and was vertically integrated with its own coal mines, fabrication plants and shipyards. In the period 1950–70 the corporation diversified into iron ore and manganese mining for export, off-shore oil production and investment in off-shore steel fabrication. And in the period from 1970 to the present the company has become a truly global minerals and energy conglomerate producing oil, exporting coal, taking over oil, coal and other mining prospects in Australia, the US and Asia, closing its shipyards and restructuring its steel division (Fagan 1987: 47).

[29] In Newcastle alone the steel workforce has dropped from 11,000 to 3,000 since the early 1980s (Maguire 1995: 9).

Australia it was and still is the subject of much analysis, particularly by leftists concerned to intervene in the processes producing regional havoc. The following discussion attempts to illustrate the discursive dominance of centeredness in both left and corporate acounts of this period and draws heavily upon the research of a number of these scholars.[30]

O'Neill (1994) has argued persuasively that the activities of BHP during the 1980s and early 1990s must be seen as a complex and contradictory interaction of, amongst other things: different and competing management strategies, flows of investment and revenue, power plays with national and state governments, discursive positionings in the local and national media, struggles between existing and future shareholders, and negotiations with workforces and communities. One way to map the contradictory processes encapsulated in the fiction of "BHP" is through an analysis of the flows of surplus value and other forms of wealth that percolated through the organization during this time.[31] The following list is divided around a crude accounting framework that distinguishes flows into and out of the company.[32]

Inward flows

Normal revenues

- Revenues from the total sales of goods and services produced in the processes of coal and iron ore mining, steel making, metal fabrication, engineering consultation and all the other production activities overseen by this complex and diverse corporation.[33]
- Large-scale borrowings on domestic and international money mar-

[30] We are deeply indebted to the research work of Phillip O'Neill, Bob Fagan, Graham Larcombe and Andrew Metcalfe for various parts of this discussion. It has been through their friendship and openness to debate over the last 15 years that many of the ideas in this chapter have been developed. We hope that any critical overtones that might enter our appropriation of their work will be taken in the comradely spirit in which they are offered.

[31] Most of these flows have been identified (though not necessarily in the theoretical terms elaborated here) and documented by O'Neill (1994: 122–215), who includes in his analysis an in-depth discussion of the ways in which they became separate foci of political struggle.

[32] Using the class analytics developed by Marx in *Capital* Volume 3 and elaborated by Resnick and Wolff (1987: 164–230) these flows could further be classified into (on the revenue side) surplus value appropriated in the exploitative capitalist class process, revenues gained as distributed class payments from other industrial capitalists external to the enterprise, and revenues gained from nonclass transactions, and (on the expenditure side) distributive class payments made to secure the conditions of existence of the exploitative class process, payments to reproduce the conditions of existence of distributed class revenues, and payments to reproduce the conditions of existence of nonclass revenues.

kets mainly to finance share buying sprees but also to aid in infrastructure improvements and introduction of new technology.
- Equity financing.

Special revenues

- State subsidies in the form of sales revenues enhanced by protection (duties on steel imports, quotas on imports from developing nations) of its near monopoly of steel production. This political intervention ensures stable and high prices for domestically consumed steel products and represents a revenue flow from the community of domestic users to the corporation.
- State subsidies in the form of infrastructure to facilitate the development of coal-mining in inland Queensland and off-shore oil and gas extraction in Bass Strait.
- Revenue flows from international industrial consumers of BHP coal and energy products which are produced under such propitious conditions (high rates of productivity and natural advantage) that when sold at prices representing international averages they are transacted way above their "value."

Windfall revenues

- Revenues gained by playing international currency markets using the vast cash flow generated by international sales of minerals and energy.
- Proceeds from the sale of steel related subsidiaries and coal mines particularly in the Newcastle region to subsidize returns in the run down local steel rod and bar products plant and thereby satisfy corporate dictates that separate steel divisions reach a standard (15 percent) level of return (O'Neill 1994: 177).
- Direct state assistance in the form of the Steel Industry Plan to "save" the steel plants from closure by guaranteeing federal government payment of $71.6 million annually for five years from 1984 to 1988. This was remitted in the form of cash bounty payments to downstream users of BHP steel (predominantly subsidiaries of the corporation) to ensure that they bought BHP products and did not allow its level of market saturation to drop below 80 percent (Button 1983: 3).

[33] This revenue would include the surplus value generated from the workers in each of these sectors.

Outward flows

Costs of production

- Wages paid to BHP workers. In the case of the mining workforce and the managerial workforce wages include a large premium to secure access to a tightly controlled labor market; in the case of steelworkers wage payments are decreasing in total (because of massive layoffs) and relatively (because of agreed upon wage freezes).
- Costs of maintaining and acquiring new plant and machinery and raw material inputs to production not internally sourced by the corporation.
- Retrenchment payments to laid off steelworkers and coal miners.

Buyouts

- Strategic buyouts of competitors in the domestic steel market and investment in new city-based mini-mills to maintain near monopoly control over domestic production.
- Buying and development of domestic and overseas production sites in the mining and energy industries to consolidate its position as an international energy MNC.
- Buying of foreign steel fabrication plants to obtain patents and brand names to enable selected Australian steel products to penetrate foreign markets.

Other payments

- Value flows to international steel users who buy surplus domestic product sold below value by BHP to obtain foreign currency.
- Huge capital flows to orchestrate highly-leveraged management buyouts of "friendly" companies in order to fight off corporate raiding by "unfriendly" ones.
- Servicing of massive debts to international lenders incurred by preventing takeover bids.
- Payments to management consultants to provide guidance on how to manage the politics of negotiating with governments and communities over industry rationalization.
- Expenditures on advertising and maintaining a large public relations division to design and promote its image as the Big Australian and to create a powerful discourse around international competitiveness as a smoke screen to hide its domestic steel strategies.
- Corporate and payroll taxes paid to the state (when not able to write

them off against depreciation allowances).

These lists make clear, or at least they make it possible to argue, that a politics of (re)distribution focused on BHP could have inserted itself among a welter of claims, none of which by itself was going to make or break the enterprise. A vision of the firm as the decentered conduit of inward and outward flows has the potential, if it were to be developed, to dislodge the more familiar image of the enterprise as a calculating subject that maximizes revenues and then distributes them in accordance with a strategic imperative that must be obeyed if the firm is to survive and prosper. But it is also important to a rethinking of distributional politics to consider how all these flows at the site of BHP were actually accounted for, that is, how various flows were positioned as important and necessary while others were seen as dependent and contingent, how their sequencing was understood as orderly and rational rather than disorderly and racked with uncertainty. Of course there have been various accounts. And each "account" interpellates subjects differently, thereby influencing thinking, strategy and action around (distributive) class politics and the project of creating different economic futures.

The corporate "account"

As it negotiated the 1980s BHP portrayed itself as a centered and unified subject situated at the periphery of an increasingly internationalized economic system. Placing itself squarely in the global arena as an "Australian" firm attempting to play it big on the world stage, BHP was able to call upon the new language of "internationalization" to present itself to the domestic gaze as an economic subject in crisis, on the verge of drastic but necessary rational action. In the early 1980s it announced that its steel production plants were no longer able to compete internationally because of aged infrastructure, overmanning and antiquated labor relations. It threatened closure of major plants unless a program of restructuring and rationalization was immediately instituted.

In popular discourse steel was seen to be the core of BHP's operations, and the company's announcement that steel production in Australia was in crisis was translated in the minds of many workers, community members and politicians alike as the danger that BHP itself was in crisis. The language of declining profits was used to support the crisis representation as "everyone knew" that profits were what kept business afloat. Presenting itself as a singular and rational subject, forced by circumstance to act to save itself and survive in an increasingly competitive world, BHP orchestrated major layoffs in the steel sector and threatened to close down its steel production facilities entirely.

Under pressure from the local communities concerned, the federal Labor government intervened "in the national interest" to save the domestic steel industry by assisting BHP in the process of restructuring its steel production plants and reshaping the labor relations in them. In addition to the millions redistributed more or less directly to BHP to assist in upgrading, the federal government via the Steel Regions Assistance Program allocated $102 million between 1983 and 1988 to assist the steel centers in diversifying their regional economies and increasing employment opportunities (O'Neill 1994: 145). Many of these funds also ended up in the coffers of the regional downstream subsidiaries of BHP.

The corporation presented itself as a unified and centered accounting unit whose rational analysis left it little economic room to move, and the federal government, bringing to bear its own sense of the centered nature of the domestic economy, went along wholeheartedly, granting support to the corporation's restructuring program with virtually no strings attached.

The counter "account"

While BHP told one story, unionists and left academic researchers told another. This story had more sub-plots, more intrigue, more complexity of character development. Replacing the transparent rational subjectivity of the corporate account was a subject (BHP) capable of covert as well as overt accounting, of self-serving two-facedness, that is. The counter-narrative goes this way: in the 1980s BHP was in the throes of internal restructuring as it sought to make the transition from a predominantly Australian-based steel and minerals producer to an international mining and energy MNC. It embarked upon a program of disinvestment in certain steel plants while at the same time using the funds generated from domestic steel sales to finance international expansion. Its highly protected near-monopoly status as Australia's only steel producer made it possible to run the industry as a domestic cash cow for its global aspirations. BHP's strong and influential political position in the national economy allowed it to bargain for massive assistance packages from the federal (as of 1983 Labor) government, using the argument that the steel plants were no longer internationally competitive. As O'Neill and Fagan point out, given that the bulk of BHP's steel sales were to the domestic market which was protected against imports, the argument about the impact of international competition was patently false. The new Labor federal government, anxious to be seen as a successful and decisive economic manager and conscious of the need to be seen as serving local working-class interests in the affected steel regions, "swallowed" the argument and acted. As one federal Labor politician put it:

> Then in '82 when the BHP thing happened and most of us had basically ignored industry. They'd [BHP] always been there and they were always going to be there, it didn't really affect us very much. Our main concern was more with the social, physical environment, density problems, road traffic, parks, all that sort of stuff: Civic Park issues . . . '82, we suddenly realised that we had a problem with our industrial infrastructure . . . So from the time I was elected in '83 I've spent most of my time, my preoccupation has been with industry. (Informant D, Metcalfe 1994: 2)

BHP's reinvestment in old steel regions "bought support" from the state for its other steel divisions and petroleum and mineral interests in Australia. At the same time it involved the state as prime negotiator of a major revision of work practices and wage structures in the steel plants. To quote the same federal Labor politician:

> We've been remaking the attitudinal, industrial side since '83, now I think people are beginning to take that for granted to some degree. They've become used to it . . . I think people actually recognise that the town [Newcastle] has grown up. A lot of them are much prouder of their companies. Their companies are putting more into marketing, into their employees. (Informant D, Metcalfe 1994: 17)

While this process of negotiation with the state was going on the corporation incurred large-scale international debts in its successful attempt to prevent takeover by the corporate raider, Bell Resources. Deregulation of financial markets in Australia allowed the destabilizing effects of raiding to jostle the fortunes of even the "Big Australian." BHP came through this period weakened in terms of its gearing ratio, strengthened in terms of its international activities, and repositioned in the domestic steel industry.

The counter "account" shows BHP working strategically at a number of different levels, using government assistance given for one purpose to serve instead or also for another. It presents a familiar picture of a manipulative and quasi-conspiratorial subject whose true character is veiled to all but certain skeptical sleuths. But it also presents the corporation as a subject riven with internal competition and contradiction. Within BHP different sections vied for investment funds and managers of product divisions fought for their own survival (O'Neill 1994). Sometimes managers found themselves identifying with their product division, sometimes with the community in which they lived, sometimes with the capitalist owners of the corporation, sometimes with the state regulators, sometimes with the finance division of the firm. The decisions they took, which directed corporate funds into some activities and not into others, were influenced by a complex notion of the total enterprise's profitability as well as by the profitability of their own divisions, their own career possibilities, and the presence or absence of community ties.

In the counter "account" the unified subjectivity of BHP (as presented in the corporate "account") is fractured into many and contradictory calculating subjects. Yet the logical structure of the counter narrative is, interestingly, similar to that of the corporate narrative. It is overwhelmingly shaped by the company's strategic pursuit of profits and the recognition on the part of government, workers and communities that an economic imperative had to be accommodated for the sake of regional and industrial survival. It is this logic that constituted the politician quoted above as "responsible" for the corporate needs of BHP ("*we* had a problem with our industrial infrastructure" [emphasis ours]) and cast workers and the local community in the minds of politicians (and corporate executives alike) as unruly children who needed to "grow up" ("I think our community is much more mature. Newcastle's grown up an enormous amount . . . " [Metcalfe 1994: 7]). Central to the orderliness of this narrative, to its logic of causality and, most importantly, to the subject positions it carved out for the company, workers, the community and politicians is the notion of profitability. That all the actions pursued were necessary to enhance and ultimately did enhance profitability is the unstated assumption that binds the events together in some sequential unity.[34]

An "other" account?

Were we less attracted to thinking being and motion in a controlled and orderly fashion, were we more drawn to representing identity as a dissipative rather than centered system, we might be able to create a very different sort of narrative around these events. In questioning the "order" of profitability we could explore questions such as: What were the effects of continuing to channel funds into the maintenance of a monopoly position in the domestic steel industry when demand was dropping, competitors were entering with mini-mills and the federal

[34] Said argues that there is an overwhelming desire for such unity in "accounts" or narratives: "Formally, the mind wants to conceive a point in either time or space that marks the beginnings of all things . . . Underlying this formal quest is an imaginative and emotional need for unity, a need to apprehend an otherwise dispersed number of circumstances and to put them in some sort of telling order, sequential, moral, or logical" (Said 1978: 41, quoted in Metcalfe 1994).

government and the corporation itself was aggressively pursuing the deregulation of markets and the rhetoric of internationalization? What was the corporate cost of strategic buyouts of competitors, of public relations and management consultants to represent itself to the public in a certain light, of bowing to the federal government's dictate that it reinvest in the old steel regions till the end of the century? What might have been the effects of allocating funds to greater international involvements, to establishing steel production in greenfield sites, to paying off steelworkers in plants that closed with wage continuation deals? And we could use such questions to highlight the contradictoriness of a corporation and its dynamics, further challenging the discourse of a unified calculating subject.

In the absence of empirical material with which to explore such destabilizing "accounts," we are tempted instead to fabricate some visions of being and motion that draw upon images of potentiality, flow and dissipation. We are tempted, that is, to an act of epistemological terrorism to present an alternative vision of BHP as decentered, fluid, disorderly and racked with uncertainty. But once again there are limits upon such a tactic. To create a discourse of the enterprise as a dissipative system, as empty, as multiple, as the site of competing and exhaustive claims, as made up of individuals who are many things (including local community dwellers and family members, philanthropists and ruthless managers, staunch unionists, and misogynists), and as a constellation of dynamics (of profitability, conservatism, disaccumulation, philanthropy, competition and protection, to name just a few) is almost too daunting in the face of an overwhelming scepticism born, perhaps, of a lack of desire for such a discourse. But while the desire to escape the power of the universal calculating subject may as yet be underdeveloped in economic analysis, there is a residual desire to think and rethink the class politics which surround events such as those described above. It is to this remnant desire that we appeal in our attempt to arouse interest in a new way of thinking economic identity and subjectivity.

Class, subjectivity, and distribution

The sequential and logical stories that were told about the flows percolating through BHP's corporate structure during the 1980s drew upon representations of centeredness, profitability, order and rational strategy, and positioned politicians, workers, the community, and the company in set roles that largely dictated their actions and responses. In particular, these narratives located the steelworkers as unitary subjects locked in class struggle with BHP, who were inherently opposed to the restructuring plan and who inevitably had to be "disciplined."[35] Their victim status as opponents in an unevenly structured power struggle dominated popular stories of regional restructuring. And in the singular representation of them as workers entitled to continued employment and deserving of a claim on social wealth via the wage payment, their "interests" were seen to be served (1) by the decision to continue steel production at the old established Newcastle and Port Kembla sites with greatly reduced numbers and new work practice agreements, and (2) by the payment of retrenchment packages to the thousands of laid off workers. From the perspective of their working-class subject position and a class politics focused upon relations of exploitation, the steelworkers "won" the right to continue to be exploited or to be temporarily compensated for not being exploited. From the perspective of the steelworkers' other class and nonclass subject positions and a class politics focused upon processes of surplus distribution, the events at Port Kembla and Newcastle were more complexly and ambiguously resolved.

For example, in their environmental subject positions as members of communities whose environmental amenity, specifically air quality, had been significantly compromised over decades by the pumping of ash and toxic materials into the atmosphere, steelworkers' and their families' interests were not served by the continuation of steel production agreed to by BHP.

In their gendered subject positions as men whose identities rested upon the positive valuation of heavy manual labor, and on an extremely virile masculinity of a definitely heterosexual nature, the decision to continue steel production with greatly reduced numbers created an interesting conflict of interest. For those workers with continued access to a steel job, the prevalent form of masculine identity did not need to be challenged. But for those workers who suddenly found themselves unemployed in towns where there were few alternative gender identities for men not linked to the masculinism of heavy industry,[36] the maintenance

[35] The counter narrative leaves no room for a multiplicity of worker subject positions as was granted, for example, to managers. Instead it hints at the possibility that, had workers had a knowledge of the corporate shenanigans taking place behind the scenes, their actions to oppose the course of rationalization might have been more effective, or perhaps more truly class conscious.

[36] Discussing the opportunity for new jobs and identities in Newcastle after the steel industry downturn, Metcalfe reports one informant as saying "A lot of people in the Hunter Valley don't like, a lot of people in Australia don't like tourism . . . They feel that you've got to run around and wait on these people, it's degrading, it's demeaning and, uh, I suppose it is in a way. But working is demeaning in a sense. Why can't you just sit there and exist like an angel? A number of people are of a fairly old-fashioned sort of attitude. Boys should do boy type jobs not go round doing poofter type jobs like being a waiter and so forth" (Informant B, 1994: 15).

of steel employment for a minority must have been a difficult pill to swallow.[37] The continued possibility, however improbable for individual workers, of still occupying the dominant masculine identity (with all the misogyny and homophobia that attend it) might well have stifled the active constitution of new gendered subject positions for men.

As class subjects within a discourse of noncapitalist economic development, that is, as potential participants in communal or collective class processes or self-employment, the laid off steelworkers were not particularly well served by the decision to pay retrenchment packages to individuals without creating any mechanisms for pooling funds to establish alternative forms of economic activity[38] or any commitment to retraining.[39]

It would seem that the class politics enacted during the 1980s in Australian steel regions was confined to maintaining access to wages for some, and ensuring the continuation of outmoded, environmentally unfriendly and masculinist capitalist production processes in order to "save the region." It left unchallenged the mono-industrial culture of the regions, the environmental and health impacts of industrial pollution and failed to secure the rights of women, minorities and future generations to any entitlement in the local economy.[40] It also left unchallenged BHP's rights to use its steel regions as a source of funds to be siphoned off through the corporate accounting system into

[37] The construction of new masculine identities associated, perhaps, with the redefinition of housework, child-rearing and neighborhood activities might have taken place had all workers been laid off. DiFazio (1985) discusses how this happened amongst longshoremen in Brooklyn in the early 1980s when they were laid off because of containerization. With the security of a union-negotiated Guaranteed Annual Income, ex-longshoremen (mostly of Italian origin) kept up an active social life amongst their former work mates and many became "family activists, conscious . . . proponents of the egalitarian family . . . no longer letting their family life happen to them but . . . actively restructuring it" (1985: 91). DiFazio sees the longshoremen he interviewed as dealing with the deprivation of meaningful work on the docks by creating meaningful work roles for themselves in their families and communities and in the process fundamentally challenging traditional sex roles, gender identities and patterns of gender domination.

[38] An attempt to create a focus for alternative economic development was, however, made at Wollongong in the early 1980s. The South Coast Development Corporation was designed to attract new industry and other economic activities to the region. Unfortunately the control exerted by BHP on the board of this organization severely limited the range and political scope of its activities.

[39] By contrast, the retrenchment agreement negotiated by the Australian Clothing and Allied Trades Union for the thousands of (predominantly female and non-English speaking) textile and clothing workers laid off after the collapse of state protection, ensured the continuance of wage payments for two years following retrenchment while providing retraining and English language classes (O'Neill et al. 1995). Within the community of workers involved in the federally funded scheme many ideas for alternative businesses are percolating.

speculative and production activities in other regions and nations.

A politics of distribution targeting the enterprise need not be disempowered by corporate and state representations of profitability as the bottom line. Economic narratives that emphasize the indeterminacy and contingencies of change, the decenteredness of economic subjectivity and the constitutive nature of politics and discourse allow for development of many alternative accounting frameworks in which different distributive flows (for instance to noncapitalist class processes, economic diversity, ecological sustainability, indigenous peoples' development,[41] or non-masculinist industrial development) and "rights" to them are privileged with legitimacy at the enterprise level. As those interested in projects of class transformation we have a lot to learn from the new languages of rights that have developed in the nonclass political arenas of feminism, environmentalism and indigenous peoples. In the face of hegemonic masculinism, developmentalism and racism, alternative rights discourses have carved out new political subject positions and different senses of justice. They have enabled people to make claims on resources and flows of wealth where previously they had none. These discourses continue to be extremely important in creating alternative emotions, desires and moralities that inspire movements of economic redistribution. Rather than continuing to suffer the personal and collective sense of pain and guilt inspired by dominant rationalist and centered economic narratives we see glimpses of a class politics of distribution that will constitute the "rights" of different and noncapitalist class processes in an economic future of diversity. While such a politics could orient itself toward the state in its familiar roles as a collection point of social wealth and as the traditional origin and agent of redistribution, it could also and innovatively focus on the capitalist enterprise as a point of condensation of appropriated surplus value, all of which will eventually be distributed. Why not to projects of economic difference and transformation, if they can establish their claims, articulate their rights, and find their subjects?

> Marx is not talking about the nongeneration of capital but the non-utilization of capital for capitalism. It's like the difference between starving and dieting. You can agree to the production of capital, but restrict it (by common consent) so that it can't be appropriated by one group of people but becomes a dynamic for social redistribution. If that kind of Marxist analysis is digested, it becomes the active core of the global grassroots movements rather than a mere model for bureaucratic state capitalism that claims a particular name. Such is the case in the new social movements of

[40] Howitt (1994a) has persuasively argued that there is never only one narrative center to restructuring and development. A view of the economic subject as multiply constituted makes this even more readily apparent.

the South – "globe-girdling" rather than international movements. They operate with the real goal of redistributing generated capital . . . This wrench between capitalism and socialism, the self and the other, between rights and responsibilities, appropriation and redistribution, taking and giving – this is extremely fundamental. (Spivak and Plotke 1995: 7–8)[42]

[41] Howitt's (1994a,b) research into the Aboriginalization of the Weipa Aborigines Society (the hybrid organization set up between the MNC bauxite and alumina producer Comalco, the Queensland Government and community members to oversee social and economic relations at the Weipa bauxite mine over 20 years ago) is a wonderful example of the way in which indigenous peoples' "rights" to determination over resource flows generated from mining on their land have informed a new kind of distributional politics focused upon a transferral from non-renewable resource development to economic and environmental sustainability.

[42] The quotation is a statement by Spivak in a dialogue with Plotke.

9

"Hewers of Cake and Drawers of Tea"[1]

In the context of discourses deriving from or influenced by political economy, including such farflung discursive regions as literary and cultural studies, we have encountered a wealth of representations of economic hegemony (presumptively capitalist) and a corresponding dearth of anti-hegemonic representations. This experience of shortage, of critical absence, has prompted us to include this chapter, which portrays noncapitalist economic activity or, more precisely, noncapitalist relations and processes of class and the interaction of these with capitalist ones. The chapter moves back and forth between an article we wrote in the early 1990s and our current – no doubt soon to be outdated – reactions to it (the latter are rendered in italics). Reflecting on some of the unintended contours and valences of this original piece, and on some of the ways in which we have since written over the same empirical terrain with different purposes and results, we present this social representation in (re)process, accompanied by a developing consciousness of itself, of some of its potential effects, and of other possibilities of rendering and being.

Writing about social existence and change we inevitably face the problem of how to represent a particular social configuration, which for us has become less a question of accuracy or fidelity (to the "truth" of what we describe or seek to understand) than one of "performativity." When we tell a story and represent a social practice or site, what kind of social world do we construct and endow with the force of representation?

[1] The phrase was used by Mitchell (1975: 9) to describe coal-miners' wives.

What are its possibilities, its mobilities and flows, its contiguities and interconnections, its permeabilities, its implications for other worlds, known or unknown? What, on the other hand, are its obduracies, its boundaries and divisions, its omissions and exclusions, its dead ends, nightmare passages, or blind alleys? How might we assess its generality or durability, its potential for transmutation? And how might its representation participate in constituting subjects of affect or action? All these questions press upon us when we consider the performative or constitutive role of social representation.

Like many pieces of writing, the piece we reflect upon here was written as an intervention in a particular discursive and temporal setting, with a particular audience in mind. Building upon our experience as students of industrial change, it represents some of the theoretical work behind an engagement with the Australian union movement around issues of industrial strategy and gender. Although the specific focus here is the impact of a new shiftwork schedule on family and community life, the broader aim is to call into question the traditional contours of industrial policy debate – the actors it legitimizes, the social sites it privileges, the boundaries that define legitimate industrial issues. This goal has led us not only to trace the effects of industrial actions on other social sites but also to attempt a more difficult theoretical task – that is, to conceptualize industry itself as an "effect" of other (nonindustrial) social practices. Thus, family and gender politics – here concentrated around the issue of shiftwork schedules – are seen as having impacts on the industrial site rather than (simply) the other way around. One of the political purposes of this theoretical intervention is to legitimate the concerns of those affected by industrial development who have heretofore not had a public voice in decisions taken by management, public officials, and unions.

But we also have a more ambitious theoretical project now, one that identifies this chapter as appropriate to a book which attempts to foster difference in economic representation. Here as elsewhere, we have drawn upon feminist work that portrays the household as a site of production and distribution as well as consumption, in order to problematize the singular representation of "the economy" as a preeminently capitalist formation located in the non-domestic sphere and unified by "the market." By portraying the economy as multiple, or as a site of difference, we are placing another nail in the coffin of the capitalist totality (if that doesn't seem too optimistic and premature). At the same time, we are specifically (re)incorporating the feminized sphere of the household into the masculinized modern economy, acknowledging the household as an economic site rather than simply as a condition of existence of "the economy" more commonly understood.

Finally, our intervention is framed in the language of class. This is a gesture toward the possibility of generating new class subject positions and interpellating alternative class subjectivities, ones that might come to life in social spaces where they are usually not theorized or permitted to exist.

Coal-mining towns have long held a fascination for social scientists and theorists, particularly Marxists. For the latter this interest is linked to the almost hallowed place coal miners have occupied in the labor histories of most industrialized nations. The image of the militant, class conscious coal miner has played a powerful role in constituting knowledges of "the working class" and "working-class struggle." And the remote mining town has been painted as home to the archetypal working-class community.[2] In the literature on mining towns and their inhabitants coal-miners' wives occupy a distinctive place. For the most part mining-town women are seen as members of the working class by virtue of their relationships to their miner–husbands and because of their hard-working participation in the reproduction of the mining workforce. But unlike the miner, whose class position defines for him a seemingly clear role in capitalist class struggles, the miner's wife's class position appears to involve a problematic relationship to working-class politics.

In studies old and new, coal miners are portrayed as committed working-class warriors while their wives are seen as political chameleons whose commitment to class politics is quite unpredictable. Sometimes painted as a conservative force, more concerned about regular income than struggles for higher wages, mining-town women are represented as a drain upon the political life blood of their more militant husbands. In contrast, at other times, especially during periods of severe industrial disputation such as protracted strikes or fights for community survival, the women are praised as the true backbone of the struggle, the militant force that keeps their husbands' commitment from wavering (Dennis et al. 1956; Long 1985; Metcalfe 1987, 1988; Stead 1987; Williams 1981).

Analyses that highlight the mercurial nature of miners' wives political behavior as it moves between quiescence and intense activism resonate with essentialist notions of women as lesser political animals – ambivalent, unconscious, naive – whose strategic vision is myopic,

[2] In Australia, the British sociological tradition of studying radical "occupational communities," exemplified by *Coal is our Life: A Study of Yorkshire Mining Community* by Dennis, Henriques and Slaughter (1956), has been upheld in the work of Walker (1945), Williams (1981) and Metcalfe (1988).

individualistic, or family centered.[3] Such views may then, in turn, actively contribute to the exclusion of women from a role in political life. In our view the construction of mining-town women as unknowable or unpredictable political subjects is in part an effect of theoretically locating them only in relationship to capitalist class processes (which effectively places them as spectators or at best reserve players in a game in which their husbands are actively involved).[4] In the following discussion we outline an alternate reading of the political activities of mining-town women. We argue that men and women are involved in a set of *noncapitalist* class processes in the household and that, for the women, this involvement often leads to participation in a set of domestic class struggles.[5] By acknowledging the existence in mining towns of noncapitalist class processes and the struggles between men and women around these processes, we may construct a different understanding of the ambivalent relationship women bear to men's political battles over their conditions of capitalist exploitation.[6]

Despite the disclaimer in note 6, it now seems quite clear to us that by making class the "entry point" we make it the center of our story and thereby tend to obscure other processes, or to subsume them by giving them life only with respect to class. Though we may not make any ontological claims for the centrality of class, we have deliberately placed it in the foreground of our social representation. Viewed with an eye to

3 As Joan Scott notes in an essay on E. P. Thompson's *The Making of the English Working Class*, women's presumed and naturalized embeddedness in the domestic sphere is often the grounds of their political stigmatization: "Domestic attachments, it seems, compromise the political consciousness even of women who work, in a way that does not happen (or is not seen as a problem) for men. Because of their domestic and reproductive functions, women are, by definition, only partial or imperfect political actors" (1988: 74).

4 This can be seen as part of the historical legacy of nineteenth-century class politics through which the categories "worker" and "woman" became effectively counterposed (see Scott 1988).

5 See chapter 3 for a discussion of class as a process involving the production, appropriation and distribution of surplus labor.

6 By choosing class as an entry point into understanding mining-town women's political behavior we are in no way wanting to undervalue other possible theoretical entry points (such as gender, psychological or place relations). We are interested in exploring ways in which class processes interact with these and other relations, not in order to see class as ultimately determining other relations, but in order to highlight class as one active process in the mutual constitution of all relations in a complex social totality. Our analysis is grounded in a concept of the "overdetermination" (see chapter 2) of any social site or process, in other words, its contradictory and complex constitution by all other processes and events. No one is ontologically privileged in the "explanation" of the object being studied, though necessarily only certain influences and interactions will come under theorization.

the performative force of language, the foregrounding of class endows it with ontological privilege – though ontology is here understood as the effect rather than the origin of the representation.

Class processes in remote Australian mining towns

The coal industry in Australia experienced a dramatic expansion over the period 1970–1988 as the industry grew to become the world's largest exporter of black coal. Production capacity expanded with the opening of many new ventures, most of which were open-cut[7] mines located in Central Queensland, at some distance from the traditional underground mining areas of New South Wales. In the Bowen Basin of Central Queensland many new "company towns" were built to house the expanded coal workforce and existing country towns were extended by the addition of suburban estates built by multinational coal companies.[8]

Some of the first "company towns" are now nearing their twentieth anniversary and the communities have matured in many ways. The physical facilities are well established. No longer are they company run; local municipal councils have taken over the administrative role. In the newer towns it is not uncommon to find two generations of miners and families (especially given the early marrying age of young people).

Permanence has not diluted the virulent "blokeland" image. The dominant presence of miners with big money, big machines and abundant machismo still pervades, although tempered by the company of many women and children whose lives are not bound up in blasting and mining coal. As the towns have aged there has been a small increase in non-mining jobs as ancillary businesses have become established and

7 Open-cut mines are those in which the coal is mined by removing the overburden (the rock and soil covering the coal measure) and excavating the coal. The mining process is closer to road building than traditional underground mining. The open-cut method, referred to in the US and UK as either strip or open-cast mining, requires an enormous up-front capital investment in earth and rock moving equipment before revenue from coal sales can be recouped. Until very recently the cost of production per ton of coal produced by open-cut methods was much less than that produced by underground methods, largely because of the lower labor input involved (now highly automated long wall mining techniques are making underground coal production in Australia once again competitive, though this has as much to do with interest rates as the comparative labor input). Open-cut mining technology has advanced dramatically since the early 1970s and many reserves have been developed the world over using this technique.

8 For more background on Australia's role in the international coal boom and the regional impacts of rapid coal developments on Central Queensland see Gibson (1990, 1991b).

local council employment has grown. Still, opportunities for female wage employment remain scarce. It is only very recently that women have been allowed access to lucrative jobs in the coal-mining production process and as yet the numbers in the industry are quite small.

Most men work full time in the mining industry. The majority are "wages workers," that is, manual laborers and tradesmen engaged in coal production and machinery maintenance who sell their labor power for a wage from the company. They can be seen as engaging in the exploitative capitalist class process of surplus labor production and appropriation in value form. A small number of men are "staff men," some of whom may have been promoted into supervisory positions from wages positions, others of whom are managers, trained technicians (geologists, mining engineers, surveyors) or office workers. Staff men manage the mining operation and supervise the labor of the wages workers in return for a salary paid out of distributed surplus value. They can be seen as participating in processes of managing the companies and supervising the productive laborers that help to secure conditions of existence for surplus value production and appropriation.

Most women work full time in domestic production, a few as casual workers who clean the single men's quarters but the majority as wives and mothers – shopping, cooking, cleaning, caring for children, organizing interaction with local service providers (teachers, nurses, doctors) and general social interaction on the part of the family. This work is unpaid and appropriated in the form of use values: meals, clean clothes, child care, an orderly, comfortable and pleasant environment. In the households of most mining families (whether headed by a wages or staff worker) it is the woman who performs the bulk of the domestic labor. What goods and services are not necessary to the woman's own survival, the man appropriates and distributes within or without the household. This arrangement is supported by ties of loyalty, the marriage vow, conceptions of the family, economic considerations, company housing policy, and sometimes by force.

The following descriptions of the gender division of labor in the household were collected by Claire Williams as part of her study of Moranbah in the early 1970s:[9]

> A man starts work at 7 o'clock in the morning and in a lot of families, the women's work is all day and all night. The majority of men don't do

[9] In this study Williams interviewed men and women in Open Cut, the fictitious name she gave to Moranbah, which at that time was a newly established company mining town housing young couples and families who had recently arrived. The responses reprinted here were given by women in answer to the question "Who has it harder in marriage, the man or the woman?"

> anything. They occasionally water the garden. Concerning the kids and the house, most men don't do anything. (1981: 147)
>
> My responsibility is to bring the kids up, he provides for them, but it's up to her to keep the house clean. His part of the marriage is just bringing home money. (1981: 146)
>
> You're always cleaning up. When he's at home, all he's doing is sleeping. (1981: 147)
>
> Women (have it harder) because they don't have the opportunity to walk out and do what they want to do, when they want to do it – have to think of the children first. When a man wants to, he knows his wife's there looking after the children. (1981: 147)
>
> Quite a bit to do, meals, washing, ironing, got to have everything ready to go away for the weekend. I'm working all the time. Men all they do is work. Men have spare time. (1981: 147)

As these comments highlight women work producing more than what is necessary for their own survival. In doing so they produce surplus labor which is appropriated by their husbands and children. The household is, thus, a site of an exploitative class process which involves the production of use values and the appropriation of surplus labor in use value form. In that the performance and appropriation of surplus labor is unmediated by market transactions and does not involve collective appropriation of surplus labor, the class process taking place can be seen as noncapitalist and noncommunal.[10] Modifying the terms of Fraad, Resnick and Wolff (1994), we identify this household class process as "feudal domestic."[11] Why it might be appropriately so identified (in terms of certain connotations of the term "feudal") will perhaps become apparent when the conditions of its existence are elaborated and explained.

In adopting the term "feudal" we hoped to draw upon some of its connotations, particularly its associations with the appropriation of surplus labor in use value form and with relations of fealty and mutual obligation. The term seemed much more evocative and compelling than

[10] In making this claim I am abstracting from the small number of households in mining towns in which communal class processes might be operating. From my interviews it would seem that a small number of families have been successful in instituting a communal household, but this is usually reliant upon the man working a fixed shift. Very few wages workers have this option.

[11] We are not arguing that a feudal domestic mode of production (complete with feudal forces of production and ideological and political superstructure) exists in these households, nor that this household is some residual form that has survived from pre-industrial days. We see these households as thoroughly modern ones where many social and even class processes take place, *one* of which is being identified as a feudal domestic class process.

the others we tried, like "traditional domestic." But a problem with "feudal" is the historical narrative with which it tends to be wedded. We are wary of depicting feudal practices as archaic or backward, with the implication that they are an evolutionary throwback lacking the robust qualities that would fit them for survival on the modern terrain. We have therefore been careful to delineate the very specific conditions of emergence of a strong feudal domestic class process on the coal fields of Central Queensland over the past 20 years, among people who may have had very different household class relations before moving to these towns.

In presenting this work to many different audiences, however, we have found them virtually united against the term "feudal." "Feudalism" carries a lot of historical baggage, as Spivak notes in her introduction to Fraad, Resnick and Wolff (1994), and we seemed unable to divest the concept of some of its meanings while retaining others. No matter that we might understand any social concept to be quite empty of meaning until its specific historical context is theorized (in this sense we see a concept or social site as "overdetermined," that is, entirely defined or constituted by its "constitutive outside"). No matter that we did not presume a commonality linking all social instances that are designated "feudal" (in other words we take an anti-essentialist approach to naming and categorization). Despite our protestations and demurrals, the practices and institutions that ostensibly characterized a particular phase of European social history seemed to stick like glue to the term "feudal."

This has given us insights into the difficulty of our more central theoretical project of divesting "capitalism" of some of its longstanding associations (for example, with dominance, natural expansiveness, wholeness, and systematicity). Certain things that we see as conditions of existence of capitalist class processes in particular historical settings (private property, for example, or imperial and colonial political relations, or environmental degradation, or hierarchical work organization) are for many people part of the very concept of capitalism. As capitalism and feudalism roll through history, it seems, they gather not only moss but trees, farms, villages and regions. Whole societies are called up by these names.

The project of divorcing the language of class from holistic social conceptions and evolutionary historical sequencing is not simple or straightforward, given the history of the use of these terms. Nevertheless it is an important step in theorizing complexly classed subjects and societies, ones where class is not the central axis of identity and antagonism and where various forms of class processes may arise and coexist. As with any theoretical and political project, the question arises whether to rework existing language or to invent new terms (in the latter case, the sense of freedom to determine meaning may be somewhat illusory, since even neologisms carry connotations). With respect to capitalism, our intent is to disturb and destabilize its prevalent meanings; generating new meanings for the familiar term is one of the ways we do this work, and this strategy requires us to use and reuse the word. With respect to feudalism, which is less central to our project, our attachment to the term is more a matter of not finding a word we liked better to do the work we had in mind.

One interesting ambiguity emerges in the text: we state quite clearly in one place that the woman is exploited by her partner (that is, he is the appropriator of her surplus labor) yet in another formulation, her surplus labor is appropriated by her husband and children. This raises a question that was first put to us by Erik Wright, who asked us how we would theorize household economic relations between parents and children. We responded, and still believe, that we could not answer such a question outside the context of a particular theoretical intervention. What becomes a matter of theoretical consequence? Why might we adopt a particular theoretical formulation? Answers to these questions fall into place only when we know what is at stake. While there is no intrinsic obstacle to representing children as exploiters, in the setting of this chapter as originally conceived, there is no rationale for doing so. Children could also be theorized in this context as recipients of gifts from their mothers (in which case these women would be operating with respect to childcare and maintenance as independent producers distributing their own surplus labor) or as recipients of surplus labor distributed by the male partner (appropriated by him from the woman) or as producers and appropriators of their own surplus labor or as producers of surplus labor which is appropriated by their parents or as some or all of these things. These theoretical distinctions and decisions did not have to be made, however, as the relation under scrutiny is the one between the woman and her partner. Theorizing children's class position(s) might be interesting but it could also be seen as superfluous and even wasteful within the discursive economy of this intervention.

Conditions of existence of a feudal domestic class process in contemporary mining-town households

In this overdeterminist theoretical setting, the feudal domestic class process is a very thin concept, verging on emaciation, until certain of the myriad processes that can be seen to constitute it are brought into the story. One of the questions that immediately presents itself is why might a rigidly gendered division of household labor, with its associated feudal relations of exploitation, come to reign so exclusively and unproblematically in the households of these new mining towns, when arguably this form of household class process is under pressure and even in crisis in other parts of Australia? It is not difficult to glimpse the specific conditions that have prompted its recent emergence.

The Bowen Basin coalfields are located some 200 kilometers inland from the east coast of Central Queensland in a region traversed by the Tropic of Capricorn. Before mining development and the construction of sealed roads, many parts of this region were inaccessible for periods of the year, especially during the monsoon season with its heavy rains. With the advent of the mining boom and the development of new open-cut mines, companies were faced with the problem of housing their workers.

Given the distance of existing centers from the new mines and the climatic conditions which could prevent regular long distance travel to and from existing townships to the minesites, many companies undertook to build "company towns" quite near to the actual mines. Workers were housed in company-built single person's quarters or, if they had families, in houses. The provision of company housing and the exercising of company housing policies is an important institutional condition of existence of the feudal form of women's exploitation in the household. A woman's access to company housing[12] is granted by virtue of her relationship to an employee of the mining company associated with each town. The lease is in the name of the employee and is valid only so long as that person has a partner (wife or de facto) and/or children living at home. For most women, having a home is dependent upon staying with their husbands:

> I know a lot of women, they get really fearful. They have to ask their husbands for everything. Everything is in their husband's name, the house is in their husband's name, it's just like they're an unpaid housekeeper. They get the housekeeping money and that's it . . . They worry because if their husband left town with somebody else, what have they got? Nothing. They can't live in their house because it's in their husband's name, they can't stay in the house unless they get a job (at the mine), they are really on their Pat Malone.[13] What can they do? They can move to Mackay or Rockhampton and get a housing commission house. (Woman, Central Queensland coalfields, 1990)[14]

[12] Which is rented at very low rates such as $10–15 per week.

[13] Australian rhyming slang for being alone.

[14] In-depth interviews were conducted during 1987, 1990, and 1991 in four different mining communities in Central Queensland: Collinsville, Moranbah, Tieri, and Moura. The four communities represent a range in age, level of company involvement in social reproduction, political history, and demographic structure. The research was funded by a Special Project Grant from the University of Sydney and a grant from the Australian Research Council.

While the housing is not the property of the miner but belongs to the company, access to housing is his right alone. A woman's access to housing is only indirectly granted. This institutional housing rule serves as a disciplining mechanism which ensures that the woman keeps the relationship with her partner viable by, amongst other things, continuing to be an acceptable provider of surplus labor.

The man, on the other hand, has to remain partnered. Should a woman leave her partner, the company has the power to evict him and send him packing to the single person's quarters. For many abandoned men, this company policy ensures that a feudal domestic class process is quickly re-established when a partnership breaks up. On the notice boards of the mining towns are advertisements for live-in housekeepers and mail-order bride brokers. The relatively large number of Filipina mail-order brides in the mining towns (most of whom are married to men much older than themselves) attests to the ease with which a recalcitrant spouse can be replaced with a "more docile" provider of surplus labor, suggesting the disciplinary force that maintains women in their servitor roles. As one woman revealed,

> My husband has told me, partly in jest, "If you leave me, I'd go and get a Filipino bride. They're supposed to be good with children." (Woman, Central Queensland coalfields, 1991)

Another important condition that promotes a feudal domestic class process in the home is the limited access to economic independence for women in mining towns. The remote mining community presents a very different living environment than that available in coastal urban areas (where some 80 percent of Australia's population lives). Many of the expectations young women brought up elsewhere might have for work, leisure and life style are precluded in these single industry towns. Couples migrating to a new mining town make a conscious decision to do so to seek work for the male partner. The lure is a good job with high pay for unskilled male workers and the promise of saving vast amounts of money. For many women this move is seen as a temporary sentence to be worked off:

> They're here in their minds to build up some money to pay off either their dream home on the coast or to get into business and get out of the mining industry altogether and live in a place of their choice. And any little bit of money that they spend that they don't see as absolutely necessary, the women, that is, not the men (the men still have their drinking money) is seen as lengthening the time they have to spend in this place. (Community worker, Central Queensland coalfields, 1990)

Unfortunately the conditions that promise wealth for men in these mining towns ensure poverty for women. Paid full-time jobs for women are very scarce; those few "good" jobs available, such as doctor's receptionist, office work at the mine, social worker or physiotherapist, are taken by the more middle-class wives of staff men, and opportunities for women to take up highly paid jobs as miners have only just begun to arise. The small number of part-time jobs available are mostly menial and unskilled such as stacking supermarket shelves, supermarket checkout, other shop assistant work, gardening, cleaning the single men's quarters, or minor clerical work. These jobs are poorly paid and usually nonunionized. The economic process of differential wage setting for men's and women's work clearly overdetermines a woman's relationship to her household. Why should she abandon her household tasks to take up some menial work for which she is paid a pittance?

The industrial successes of the Combined Mining Unions in winning wage increases for their members over the period of the long boom have contributed to women's entrenched position in the home. In towns where the male workforce is solidly union and where mineworkers' wage packages are in the range of $30,000 to $70,000 per annum, the wages most wives could earn in a week could be made by a husband in one overtime shift:

> I don't think the women have the need to work now, though, I mean the married women . . . you don't need to with your wages. I mean John has only got to work another doubler a week and he'd get as much as you would in a casual job, . . . that's if the tax man doesn't get it all. (Woman, Central Queensland coalfields, 1990)

This economic disparity partly relates to the unevenness in union representation across industries. The success of the mining unions in winning wage increases and improvements in working and town living conditions is starkly contrasted to the lack of union representation and industrial muscle of workers in retail outlets, cleaning jobs and other sites of female employment.

The greater the disparity between a man's and woman's wage packet, the less likely it is that female employment will be deemed worthwhile, especially as jobs in these towns are often seen primarily in their economic and not their social context. Married women seeking work are not viewed positively as wanting some independence, or as eager for the social interaction a job, no matter how menial, provides, but rather are disparaged as being greedy for more money. These social judgments on women working outside the household are enough to discourage many from paid work.

While institutional and economic factors provide important influences that promote a feudal domestic class process in the home, conceptions of masculinity and femininity are also active constituents. Mutually exclusive binary gender identities are continually under construction in mining communities, where women are seen as naturally the primary carers and nurturers and men as the rightful decisionmakers and providers. Until very recently mine work has been seen as "men's work" – heavy, dirty and dangerous – although since women have been allowed to take up mining jobs (against strong opposition from many men in the industry) this image has been actively challenged.[15] Despite this victory for equal opportunity, and perhaps because of it, "traditional" ideas of gender identity abound.

It is interesting to consider whether we intend the term "traditional" to denote longstanding and prevalent gender roles as opposed to ones that are relatively recently instituted or locally naturalized as hegemonic. We have usually been inclined to theorize binary gender – in its extreme mutually exclusive form – as itself quite a fragile and rare formation, one that only arises as a social phenomenon in "extraordinary" circumstances like the single industry town. So while discursive examples of mutually exclusive binary gendering might be quite common, such genderings are only infrequently and temporarily embodied. In this way we are able to represent the social space as a space of gender diversity and overlap, while acknowledging the existence and even the dominance of mutually exclusive gender in the discursive realm.

Bea Campbell (1986) has argued that, in mining communities, gender in the form of the division of social labor into (mutually exclusive) male and female roles was an outgrowth of nineteenth century political struggles that transformed coal mining from a community-based industry involving male and female workers into an exclusively male preserve. This exclusive and exclusionary form of gendering is now a real barrier to female participation in a mining industry in which the labor process and the legal context (including affirmative action regulations) have been substantially transformed (Eveline, 1993). It is also a barrier to male participation in certain household economic activities; even if they are retired or laid off, many men find it difficult to countenance the regendering that goes along with taking on particular household tasks,

[15] Miners' wives have an ambivalent relationship to the few women miners who have managed to get jobs with their husbands. Many mentioned they didn't think it was appropriate work for females, or that the ones who took it up were "big and burly" (not feminine as women should be). In contrast, one of the women miners I interviewed claimed that most women she knew would, in fact, "give an arm and a leg to have the opportunity of a mine job and the independence that gave."

and many women are reluctant to give up those tasks for similar reasons.

Homosociality is another ingredient of gendering in mining communities where men and women tend to socialize in same sex groups (even when they come together as families). Although homosociality is not necessarily associated with any particular form of sexuality (Sedgwick 1985), in these mining communities it affirms strictly defined masculinity, femininity and heterosexuality.

This is not always the case in mining towns, especially remote ones. Sometimes male miners who live away from women for long periods of time may engage in acknowledged sexual and domestic relations with other men, without losing their heterosexual identity or experiencing a regendering in the direction of feminization. In the South African gold mines of the earlier part of this century (extending through the 1950s in some cases) black migrant workers frequently took "boys" as domestic and sexual partners, who became widely known as "wives of the mine" (Moodie 1990: 413).[16] *The younger men took the traditional female role in a strict division of domestic labor and sexual activity (they were not permitted to ejaculate "between the thighs," for example).*[17] *This feminized them while masculinizing their senior partners, who became more manly by virtue of having taken a "boy."*

The domestic arrangements of these "mine marriages" often involved a feudal domestic class process in which the "wife" provided services to be appropriated by the miner "husband" or "boss boy," usually in return for presents or a share of the miner's wages. Obviously the feudal household itself is not incompatible with a same sex domestic partnership, or with homosexuality (and neither of these things entails the absence of male-female gendering). For the "boys," however, the outcome of gendering was a multiple or ambiguous gender identity, since they were identified at one and the same time as young men and as wives. Moreover, the purpose of becoming a "wife" was to earn money to purchase cattle, pay for a wife and build a homestead, thereby acceding to full manhood.

For both the migrant miner and the "wife," the mine marriage was part of a strategy of avoiding or resisting proletarianization. It gave the younger man access to the cash that was needed to establish an independent livelihood on a homestead. For the miner, it provided an alternative to "town women" who might divert him from the project of building a

[16] I am grateful to Glen Elder for drawing my attention to this point and to the work of Moodie.

[17] A traditional form of nonmarital sexual activity among the tribesmen being interviewed.

homestead and draw him instead into a permanent relationship near the mine, probably involving children. The mine marriage, by contrast, was temporary and entirely consonant with returning unencumbered to his home place, leaving the mines and the life of wage labor.

The choice to migrate to a mining community is a life-style decision in which the influence of a discourse of traditional gender roles in the social division of labor is strongly felt.[18] For many families, migrating provides the opportunity to afford raising children without the mother being forced by economic necessity to seek wage employment, as she would be in most urban centers of Australia. For women whose primary self-identification is with being a wife and mother, the move to a modern mining town is an act of liberation from the economic constraints placed upon such a role elsewhere. For others, the sojourn in a mining town is the result of a decision to (temporarily) abandon their work/career and to devote themselves full time to mothering, household duties and saving money (Sturmey 1989).

Central to a traditional view of being a wife and mother is acceptance of the notion of service to the family. This means that the performance of surplus labor in the household is usually not resented, but is viewed as a natural part of the role of wife and mother. Most women are reluctant to find part-time employment outside the home (and some limited form of economic independence), unless this work can accommodate the husband's pattern of shiftwork which rotates on a weekly basis from afternoon shift (3 p.m. to 11 p.m.) to day shift (7 a.m. to 3 p.m.) to night shift (11 p.m. to 7 a.m.). Thus their belief in service to husbands and children involves women in a complicated and tiring work schedule:

> When he's on night shift I wait up till 10:30 p.m. till he goes off. Then on afternoon shift I stay up till he gets home – it's after midnight when I get to bed. But then I've got to get up with the kids at 7 a.m. (Woman, Central Queensland coalfields, 1990)

> When my husband is on day shift (7 a.m.–3 p.m.) I get up a 5:30 a.m. every morning to prepare him a cooked breakfast. Then I go back to bed until the kids wake up. (Woman, Central Queensland coalfields, 1991)

> When I first came here I was surprised to hear women talking about being on afternoon shift or day shift and having to get home or not

[18] It is interesting to note that in Williams's sample of 52 couples interviewed in Moranbah in the early 1970s, only one-third said that they had jointly made the decision to move. In the rest of the cases it had been wholly the husband's decision (1981: 149). This suggests that couples attracted to the mining town might have been more likely to have had a very traditional marriage, perhaps involving a feudal domestic class process, prior to migration.

being able to make it to a meeting. I assumed they must have a job but I soon found out that they were talking about their husband's shift. They identified completely with their husband's work pattern and if he was on afternoon shift they would have to get home early to make the lunch for him before he went off to start work at 3 p.m. (Professional woman, Central Queensland coalfields, 1990)

In the unique conditions of these mining communities acceptance of a role of servitude in marriage binds women even more tightly into a feudal relationship with their husbands. But fitting into this tiring work schedule is not only seen as drudgery by the women; it is also one way in which they can glean some companionship from their relationship:

It's hard when he's on afternoon shift. The movie on TV usually ends at 10:30 p.m. and then I try and wait up till 11:30 p.m. to see my husband. It's a really nice quiet time, no kids around. You've had time to calm down after the evening war of dinner and bedtime and he usually takes a long time to unwind after afternoon shift. But then I don't get to bed till 1 or 1:30 a.m. and I'm up again with the kids in the morning. By the end of seven days of afternoon shift I'm absolutely exhausted. (Woman, Central Queensland coalfields, 1991)

The experience of being a wife is accompanied, and indeed made tolerable and pleasurable, by notions of sharing and partnership and, for many women, the sense of power they gain over their environment. One job which is largely in their hands is the planning, organizing and coordinating (often referred as having the responsibility or worrying) of the household. Women act as the managers of domestic production, they act as supervisors of (their own) labor, pacing the production process and monitoring quality control. Another job frequently performed by women is the management of the family finances. In her study Williams (1981: 154) found that the notion of a "common purse" was frequently mentioned by Open Cut women. This is the name given to the husband's wage to which the woman has access so that she can make all of the monetary payments associated with the household:

She has the responsibility, children, everything. Her husband gives her the money and she pays for everything. (1981: 147)

Man never has to worry about money, she makes sure the bills are paid and makes sure all the family got clothes, food in the cupboard, money for petrol. My husband and father are the same, they don't carry a wallet. They have the responsibility of bringing in the money but he hasn't got the responsibility of buying clothes. (1981: 147)

In performing the work of labor management and financial management women engage in distributing some of their surplus labor to ensure that domestic reproduction progresses smoothly. They thereby work to secure the conditions of existence of their own feudal domestic class position (Fraad, Resnick and Wolff 1994).

By naming the family income the "common purse," rather than the man's wage, an image of equality and family sharing is manufactured. No longer is this money seen as a payment to the man over which he has total right of disposal. As the "common purse" it is seen as a sum over which women have equal rights. The sense of power gained from this discourse of equal access,[19] and from the unquestioned role as family manager, is one of the noneconomic conditions that make the feudal class process workable.

Most relationships are accompanied by love, emotional support, and a host of companionship practices that reinforce and stabilize feudal domestic class processes in the home. But alongside these emotionally satisfying forms of enforcement are other more frightening practices. Domestic violence is common amongst coalfields families, as it is in many city families, but escape for women and children here is even more difficult. The nearest refuge is quite a distance away, either in Emerald or on the coast. In these remote communities the relationship stakes are high. Leaving an intolerable situation is not easy for a woman:

> I've heard a lot of women say "Why should I leave and let that bastard have it all? I've had his two children. If I leave here, where do I go? What do I get? I've got nothing." (Woman, Central Queensland coalfields, 1990)

Legislation now exists to make domestic violence a chargeable offence, but the placing of an interim order on a violent man by police is dependent upon the woman, or a witness, making a statement to local police at the time or soon after a violent attack while evidence is still available. Most women involved, or neighbors who witness violence, are unwilling to come forward.

[19] Although Williams found that many of her sample had joint checking accounts and that women performed all of the financial management for the family, this did not mean that they necessarily had an equal share of economic power in the household. She heard stories of unhappy wives leaving town with their kids only to find, when they reached the coast, that access to their joint accounts had been cut off (1981: 153).

In my own interviews it became clear that some women learned of their partner's increased wages only with the arrival of a new car and greatly increased deductions from the weekly income for car payments. So the control women may feel they have over the family wage is not always respected by their husbands.

If the impulse behind this intervention were to portray difference in the domestic realm, this might be an appropriate place for the invocation of ethnic difference. The mail-order Filipina brides have a particular relation to domestic violence that derives from their membership in a foreign minority group with certain solidary expectations. In one case a young bride who was planning to escape to the coast was visited by the local Filipino community and enjoined to stay with her husband or risk giving Filipina brides in the region a bad name.

Whereas staff men might subscribe to the belief that their wives should not be tied to the home and therefore support activities outside the household, mine workers appear to have a different view. Only certain types of women's activities outside the home are condoned. These are activities to do with sport (fitness) and children. As Williams also discovered, workers' wives rarely belong to clubs for their own self-expression (1981: 145):

> I'm a stayer at homer. I don't go anywhere. I'm prepared to stay at home and do my own thing around here. (Worker's wife, Central Queensland coalfields, 1990)
>
> I was secretary of the netball club, I used to play netball, go to aerobics. The kids were in the swimming club and the football club. (Worker's wife, now a miner, Central Queensland coalfields, 1990)

In many respects the feudal domestic class process in workers' households is less hidden by involvement in outside activities than in management households. And attempts to create conditions under which feudal exploitation in the home could be reduced have been fraught with problems. For example, when workers' wives attempted to get union support for the establishment of a child-care center in one of the mining towns, the men refused to allocate any funds. At a public meeting (in the shower block at the minesite) it was made clear to the women that miners would not support anything that smacked of feminism, or that would undermine the work of their women and/or allow them to start thinking about leaving home. As one unionist put it after the center had been built:

> Don't let your wife go there (to the child care/community center) unless you want her to divorce you. (Male unionist, Central Queensland coalfields, 1987)

By refusing to support the day-care center the men were implicitly refusing their women the right to child-free time to pursue their own interests, let alone any work possibility that might arise. Not surprisingly, the use made of the child care center by workers' wives is minimal:

> They don't see the day-care center as something they can use legitimately when they are not working themselves. It comes down to a question of money. The workers' wives stay at home and go to aerobics. They'll play sports an awful lot, it's a really big thing in town. They get involved in their kids things, so they go and serve with the Parents and Citizens Association at the school . . . the kindergarten committee. They tend mostly to stay at home and be with the kids. (Community worker, Central Queensland coalfields, 1990)

The preceding discussion has explored only some of the many interacting conditions which converge in the households of mining towns to constitute a feudal domestic class process. Company housing policies, lack of women's employment, men's work schedules, macho images and traditional views on family roles all conspire to promote a form of class process in the home which, elsewhere in Australia, may be on the wane or in the midst of transformation under the influence of the women's movement or the exigencies of the labor market. In these isolated mining-town environments women are engaged in a feudal domestic class process in the household which is very different from the capitalist class process their husbands engage in at the minesite. At times the operations of these two class processes mesh quite easily in a mutually reinforcing relationship. At other times the meshing jars, as the following analysis suggests.

We are aware that many people may see the existence of these households as a problem in itself. When we have given talks about domestic life and women's work in coal-mining towns, our listeners often remark that life there must be terrible for women. Our interactions with these women, however, did not uphold this impression. The women did not seem particularly unhappy or unfulfilled. Even many of their gripes about their husbands seemed to be acceptable slurs slung across the gender line, rather than expressions of deepseated animosity or dissatisfaction. Given the hours that men worked, including their commuting, and the porousness of the women's working day, the gender division of labor and the household exploitation it entailed for women did not seem to them or to us to be particularly unfair. Perhaps for these reasons we did not depict the feudal household as something that should necessarily be undermined or abandoned, despite the things we saw in it that we didn't like, and despite our interest in promoting communal class processes. For it is one thing to be interested in promoting and allowing for difference – in the practices of sexuality, gendering, and household class relations, let's say – and quite another to suggest that a particular form of these practices should be eradicated entirely.

If we suggest, however, that under certain circumstances exploitation may be quite acceptable and even desirable to the exploited, we may expect that to change when those circumstances change. What we find in the discussion that follows is just such a change in circumstances, one that made household exploitation unpleasant, unfair, and even intolerable to many women in these coal-mining towns.

Contradiction and change

In the late 1980s the coal industry entered a crisis; over-production at the international scale had led to declining coal prices and market downturn. In Australia underground mines closed, miners were retrenched from even the most modern open-cut mines and coal companies were bought and sold (Gibson 1990). The declining world market and low company returns strengthened the employers' resolve to push through a new Industry Award designed to "remove restrictive work practices" and increase production levels at lower cost. With employment levels declining the Combined Mining Unions were forced, largely against their will, to accept the ruling of the Coal Industry Tribunal (the arbitration court for the black coal industry) and new work practices were instituted in 1988. Aspects of these new work practices have had a major effect upon miners' households and the operation of feudal domestic class processes in the home.

As part of the move towards greater "flexibility" (for the companies), the new award involved widespread adoption of a new work roster called the seven-day roster. This roster involves dividing the workforce into four groups such that at any point in time there is one group on each of the day, afternoon or night shifts and one group off work. Workers labor for eight hours in each shift for seven consecutive days, afternoons or nights which advance on a rotation from afternoon to day to night. Between each shift workers have a short period of one or two days off, and at the end of the cycle a longer period of four days off. For the companies this roster has the benefit of maximizing production output and minimizing manning levels. It was sold to the workers because it maximized income as it allowed the most weekend (involving double payment) days to be worked within the month.

In the campaign leading up to the introduction of the new roster, companies in some towns sent letters to the miners' homes outlining the wage increases that would ensue, hoping to enlist the support of what were perceived to be the "money hungry" miners' wives.[20] While many women did not like what they read, there was no concerted opposition[21] to the roster as most men were indeed enthusiastic about the promised increase in wages:

> The guys had dollar signs in their eyes. (Unionist, Central Queensland coalfields, 1990)

> The other day an 18-year-old school leaver walked into a $60,000 job – he's on the seven-day roster. (Unionist, Central Queensland coalfields, 1990)

At most mines workers accepted this assault on the effective length of the working week and on employment levels in return for the jump in real wages that the package delivered (for many miners the increase was not insubstantial, ranging from $10,000 to $30,000 per year).

The effect of this decision on the community and, of course, upon women has been great. For many, the domestic work of women has risen and established companionship practices of mothers, fathers and children have been largely destroyed. Although this roster incorporates the longest break between the end of one roster cycle and the beginning of the next (from 7 a.m. Friday to 3 p.m. Wednesday) it allows only one consecutive Saturday and Sunday off in four:

> The mother's got to take on the father's role on the weekend. Men are working harder and so are women. (Union delegate, Central Queensland coalfields, 1990)

> I don't think the women are involved enough in what's going on in the industry. I mean, they are the ones that have got to be at home. When your husband's on shiftwork, you're the mum and dad to both kids, especially people on seven-day roster. You know, the kids have got football out at Middlemount on the weekend – it's the mum that's taking them out and stands screaming on the sidelines when it's Dad that should be there to see Johnny – Dad's at work. All to earn this money . . . for what? Three or four years down the track, well, mother's going to say "I'm packing up." (Woman, Central Queensland coalfields, 1990)

[20] In Collinsville this tactic was met with a strike. The union representatives argued that if the company wanted to conduct its industrial negotiations in homes then the miners needed to stay at home for a day per letter to read the mail and discuss its contents with the family. This soon put an end to this strategy at Collinsville and this is the only community in Queensland where the introduction of the seven-day roster has been, until very recently, successfully opposed.

[21] Unlike in New South Wales, where the wives of miners at a mine in the Hunter Valley made submissions to the Coal Industry Tribunal opposing the seven-day, four-panel roster.

This new roster places an extra burden upon women in the household. The remote possibility of being able to take on a job outside the home is now significantly reduced. If there are children at school, the mother is effectively a solo parent three weekends out of four.[22] The feudal domestic class process in the household has begun to experience significant stress. Already the seven-day roster is being called the "divorce roster":

> The seven-day roster was one of the things that killed my marriage. (Woman, Central Queensland coalfields, 1990)

> My husband is usually on permanent day shift, but he just had to relieve someone on a seven-day roster for three weeks. We hated it – we were ready to divorce after it. We didn't see each other to talk to for three weeks! (Woman, Central Queensland coalfields, 1990)

> Before seven-day roster my husband did certain jobs around the house – all the yard work, because he enjoyed that, he always did the washing up and so on. But now I just can't depend on him. When he's on afternoons (3 p.m.–11 p.m.) or nights (11 p.m.–7 a.m.) he's away or asleep and can't do the washing up, so I do it. He can never remember when to put the garbage out now. So I tend to do everything now with no help. Also, he's so buggered I try and save him from having to do anything around the house. (Woman, Central Queensland coalfields, 1991)

There are no figures on family breakdown since the introduction of the new roster system, just the hearsay and personal experiences of coal town people (Pearce 1990). From anecdotal evidence, it would appear that most families found that feudal domestic class relations in the home were viable when underpinned by the five-day rotating roster and voluntary overtime, but now, with an imposed seven-day rotating roster, the glue which kept many families together is coming unstuck.

Elements of the unspoken feudal contract are now being called into question as men can no longer participate in family life as they once did. In some instances they are no longer providing the protection that is expected of them:

> The worst shift is night shift. Before, he only rarely had to work the night shift. Night shift is the killer. Still we're not too bad now 'cause I've got two dogs around here. So I'm not as scared as I was. We've put up a six-foot fence and got two dogs. (Woman, Central Queensland coalfields, 1990)

For many families it is the effect on social life that is most noticeable. Much of the social life of these isolated towns revolves around weekend

[22] For families with pre-schoolers the effects may not be as drastic.

activities – sports, barbecues, dances. Now half the mining workforce works either day or afternoon shifts during the weekend:

> We have friends out at the railway houses, near the mine. Now we used to see them possibly every second weekend – they could come in here. My husband used to work with him on the three shifts. But now with the seven-day roster he doesn't get to see him as much so, therefore, we don't get to see them as much, as we used to know what shifts they were on. We used to have card nights. Now we can't do that. We drive out there in the morning when my husband's on afternoon shift, but only half the time Joe's not there he's at work on day shift so we get to see Carol. Never the four of us are together all the time. If you want to have a party one can't stay long because they're on day shift. That's what really puts the pressure on. We used to go to dances and everything like that, but we can't do that anymore. (Woman, Central Queensland coalfields, 1990)

Men and women in mining towns see themselves as engaged in a joint project working toward certain goals – a good upbringing for their children, a house of their own on the coast, a comfortable retirement or a different life, perhaps in a small business, after savings have been accumulated. Commitment to this joint project (and the feudal domestic class process it promotes) is sustained by a discourse of love and companionship between partners.[23] The seven-day roster has wreaked havoc with established companionship practices. The weekdays off between shifts and the long break between roster cycles allow men and their non-working wives to see each other – but at times that do not coincide with children's or other friends' time off. This time may also coincide with the few activities the worker's wife participates in out of the home – tennis or play group – so it creates a conflict for her. Does she stop all social interaction in the community and sit at home and watch a video with her husband, or does she leave him alone, perhaps to go off to the pub for a solitary drink?

The seven-day roster has stripped away the activities and notions of the family which were important conditions of existence of feudal domestic class processes in the home. The man has been reduced to his economic function alone. For men and women alike this situation is engendering a domestic crisis:

> You tend to feel you're living there (at the mine) now. (Miner, Central Queensland coalfields, 1990)

[23] Williams found in her interviews with 51 couples in Open Cut in the early 1970s that "the companionship of doing things together" was the most valued aspect of marriage by both husbands and wives, though wives valued it more strongly than the men (1981: 136).

I've told the company about the effects on the community, sporting clubs, but they just say they don't care, they are here only to mine coal. (Miner and unionist, Central Queensland coalfields, 1990)

The seven-day roster? I loathe it. Men were not meant to work seven days in a row . . . they get so ratty. (Woman, Central Queensland coalfields, 1990)

They (the men) knew what to expect but they didn't know how hard it was going to be. (Woman, Central Queensland coalfields, 1990)

Companies, too, have begun to feel the negative effects of the roster in that absenteeism is on the rise.[24] It is not uncommon for men to take a sick day (usually referred to as a "sickie" in Australia) when on the long stretch of night shift, or on a weekend when a particular sporting event is on. At one mine a particularly unenlightened manager attempted to recruit miners' partners in an attempt to reduce absenteeism. Assuming that wives were preoccupied with the pay packet and would side with the company and want to discourage their men from taking illegitimate sick days (for which pay would be withheld), the company instituted a new rule necessitating, in addition to a doctor's certificate, a letter from the sick miner's wife verifying his condition. But as more women suffer the day-to-day consequences of the seven-day roster, women cannot be relied upon to police the conditions of capitalist exploitation (and ensure the smooth flow of income into the feudal household). Increasingly there is an incentive for them to encourage absenteeism on the part of their husbands. The women have realized that the noneconomic male support role is as important in the feudal marriage as is the economic role of bringing in the money:

Some of these ladies now are saying they're not going to sign these letters. "Bugger 'em" they're saying, and I don't blame them. I think it's a great thing. But as a union official I've got to abide by the Coal Reference Tribunal. Some of the women are starting to get a bit hot underneath the collar. I don't blame them. I would not like my wife to sign one. Matter of fact I'd prefer to go without pay. I'd prefer to say "Stick your sickie up your bum, chum." (Unionist, Central Queensland coalfields, 1990)

The introduction of new rosters has produced contradictions in the operations of feudal domestic class processes and has seen the beginnings of domestic struggles over the increased rate of feudal exploitation in the household. In some cases these struggles have resulted in family

[24] Since the introduction of the seven-day roster in the Queensland mines absenteeism has risen by 11 percent across all the mines in the state (Queensland Coal Board *Annual Report* 1990: table 8.8).

breakdown or divorce, in others they have taken the form of absenteeism by men from mine work. In still others struggle is being waged in various emotional forms.

In Central Queensland no public opposition to the seven-day roster was organized by miners' wives. By contrast, in the Hunter Valley, an older established coal-mining region in New South Wales, women successfully organized opposition to its institution on the grounds of its incompatibility with community and family life. They made industrial history by becoming the first group of women to address the Coal Industry Tribunal and were celebrated as true working-class heroines when the issue was settled out of court with the institution of a more socially acceptable roster system.

It might seem that the Hunter Valley women recognized their class interests and resisted their gender subordination, while the women of Central Queensland were unable to step out of their subordinated gender roles and pursue the interests of their gender and class. But perhaps this reading is an effect of a familiar conception of class politics as a public and solidary process organized in a military fashion around interests ordained by an economic structure. It is interesting to think about the possibility that other forms of political action and subjectivity could be acknowledged or engendered through a different discourse and conception of class (Gibson-Graham 1995a).

The lack of any public *political response by women to the change in work practices need not necessarily be read as a lack of interest or as willing disengagement from class politics. Indeed, these Central Queensland women may be seen as actively involved in struggles around their own exploitation that are largely hidden from sight, enacted in the emotional and domestic arena of mining-town life. What is more, these class struggles have an impact upon class conflict conducted in the public arena of the mine site and industrial arbitration court.*

Given the usual isolation of the feudal housewife laboring alone in the Central Queensland household, her class struggles are almost always waged alone. Many influences undermine any sustained basis for solidarity amongst feudal producers in Queensland coal mining towns. Women possess a clear consciousness of their dependence upon their male partner. They know that protection and livelihood are linked to maintaining a relationship of loyalty to him and making sure he does likewise. "Keeping your man," that is, preventing other women's access to him and his to them, becomes a preoccupation with an added edge of desperation given the housing rules already outlined. For many women this leads to a rather lonely existence which does not lend itself to building links with others in similar situations, as these comments made by women imply:

> I have no intimate buddies here. (Woman, Central Queensland coalfields, 1990)

> I keep to myself a lot, I don't like mixing with other women much, they all talk about gossip; who's leaving with who, family break-ups. (Woman, Central Queensland coalfields, 1990)

> I used to play a lot of sport, but the women are so rough and bitchy. I'm pretty small and sometimes I was scared of getting hurt by some of them. (Woman, Central Queensland coalfields, 1990)

> A lot of women do aerobics, but that's mainly because their husbands are saying "You're looking fat." (Woman, Central Queensland coalfields, 1990)

The fear of being replaced is an underlying tension which can translate into competition and separation from other women rather than a recognition of sisterhood and a common cause.

But this does not mean that workers' wives are not engaged in class struggles in the home on an individual basis. The introduction of the new roster system has promoted an increase in individually organized class activity. In response to the intensification of women's labor the seven-day roster has imposed, some women have opted to leave their partners on the coalfields and re establish households elsewhere based upon an independent class process. In these sole-parent households women produce, appropriate and distribute their own surplus labor. Their abandoned men are forced to rely upon other class processes to supply domestic services in the single person's quarters. Divorce and family breakdown can, in this situation, be seen as the outcome of class struggle.

Andrew Metcalfe (1991) has recently argued for a reconsideration of the use of the metaphor of war which is so prevalent in established Marxist theorizing and strategizing about class struggle. He suggests that the metaphor of divorce may be more useful or appropriate as a conceptual tool. We take this one step further to suggest that divorce itself could under some circumstances be seen as class struggle.

In a later phase of this research we attempted to address the isolation of the Queensland women by involving them in an action research project focused on the family and community effects of the seven-day roster. We brought together twelve women from four towns for two two-day workshops to prepare for and then process the results of in-depth interviews they each undertook with six other women in their towns. Over the lengthy course of this research we have come to see ourselves as engaged in a political project of discursive production and destabilization. With the women who became our co-researchers, and through the process of engaging in collective research, we generated a new public discourse of mine shiftwork and a new subject position of "mine shiftworker's wife". The discourse of mine shiftwork and its impacts on the household economy and community life has begun to circulate in published form, in a booklet entitled Different Merry-Go-Rounds: families, communities and the 7-day roster *(Gibson 1993), published and distributed to union and community members by the mineworkers' union. Within this discourse the "mine shiftworker's wife" is a subject position that has given some women a basis of partial identification with other mining-town women, not on the grounds of their experiences as "women" nor of their supposed status as "working class" but on the basis of a shared, externally related identity as partners of shiftworkers affected by the rosters. This "partial identification" has helped strengthen the women's public voices and provided a shared and legitimate position from which to express their concerns in an arena from which they had long been excluded – the historically bounded and demarcated sphere of industry policy and politics. As women began to inhabit the subject position of "mine shiftworker's wife," they actively displaced the existing discourses of "mining-town women" and "miner's wife" that had constructed them as politically ambivalent or backward subjects.*[25]

It is interesting to speculate about the other forms of class struggle and transformation that are taking place at present. Some women have attempted to resist intensification by encouraging their partners in absenteeism (which thus maintains the existing organization of feudal domestic class processes). Others have been able to convince their partners to redistribute a portion of the increased wages into subcontracting parts of the household production process (cleaning, baby-sitting) out to self-employed contractors. In this case independent commodity production would be introduced into the household and women would take on the additional work of engaging an independent producer as well as playing their own role in a feudal domestic class process. Still others are continuing to adjust to the changed situation and experiencing the exhaustion and emotional drain that intensification of labor entails. At present, there are clearly many constraints upon mining-town families adopting communal class processes, whereby men and women participate

[25] For an extended discussion of this research process and its outcomes, see Gibson-Graham (1994b).

jointly in the production, appropriation, and distribution of surplus labor in the household.

Conclusions

This discussion of the seven-day roster highlights some of the interesting contradictions that have emerged from the most recent episode of industrial restructuring in the coal industry. Despite the jubilant predictions of increased productivity proclaimed by the mining companies after the 1988 Industrial Agreement, many mining operations were not to see the increases their managements expected. In the early 1990s some companies abandoned the seven-day roster and reverted to pre-1988 work schedules, while others have continued to promote it because it signifies an ideological victory over the mining working class.

On the basis of the class analysis developed above it would be difficult to see the contradictions associated with new work practices in the coal industry as arising only from class struggles over the production and distribution of capitalist surplus value. On the Central Queensland coalfields an important condition of existence of capitalist mining operations is the feudal household. The changed work practices introduced in the coal industry have had major consequences for both men and women living in feudal households. The induced changes have not been supported by women despite the increased male wages they have produced. Instead, these changes have instigated household crisis, the results of which have fed back to affect the coal industry. In this way the conditions in households and women working within them that have previously operated to secure the conditions of capitalist exploitation have more recently begun to undermine them.

Usually only capitalist class processes associated with the coal industry are recognized as structuring life and struggles in mining towns. We have argued for the representation of feudal class processes associated with household production as another important axis of social organization and struggle. Men's participation in both capitalist and feudal domestic class processes, and women's almost exclusive participation in feudal ones, overdetermines the political roles women are able to take in mining communities. Women's reluctance to engage with their husbands in union struggles with employers cannot be seen as an abrogation of their working-class responsibilities, or an expression of a "false consciousness." The so-called political ambivalence of women, compared to the class militancy of men in mining towns, can be seen as a misreading of the different and often contradictory relationships of women and men to capitalist class struggles. The isolation and privacy of domestic class struggles keeps the class politics that many women are engaged in hidden from view.

Interestingly, it has been during the recent period of industry work practice restructuring that the contradictory politics around capitalist and feudal domestic class processes have generated crucial insights into the workings of the miners' households. The many noneconomic conditions of existence of the class process in these households have come under assault as the seven-day roster has concentrated men's leisure largely into weekdays rather than weekend days. The increase in capitalist exploitation at the mine site has been paralleled by an increase in feudal exploitation in the household – and women are fighting against this increase, perhaps more concertedly than are their partners. The effects of women's resistance to greater exploitation are being felt in the coal industry as absenteeism rises.

There is some evidence that coal companies are unhappy with work rosters as they are currently organized.[26] Company management and union officials are considering alternatives that will reduce absenteeism and yet not decrease wage packets. Some companies have introduced 12-hour shifts and others have imported an "expert" from the United States to advise them about the physiological and psychological impacts upon workers of different shift systems.[27]

Awareness of the significant impact industry restructuring is having upon household restructuring is growing. It would appear that gradually miners are beginning to see the connections between their work patterns and those of their wives. The seven-day roster is commonly referred to as the "divorce roster," a tacit recognition of the domestic crises taking place as feudal households are ruptured and men are left to become independent domestic producers, or occupants of the company

[26] The additional wages bill involved with the increase of weekend work has been too great for some mines to meet, causing the seven-day roster to be abandoned. At other mines, companies have increased the use of capital equipment without allowing for adequate maintenance time and this has led to an unacceptable rate of mechanical breakdown.

[27] In yet another typical example of "cultural cringe," companies and unions have paid hundreds of thousands of dollars to import the expertise of Dr Richard Coleman, author of *Wide Awake at 3 a.m.: By Choice or By Chance?*, to devise more workable shift systems. In the opinion of psychology-trained Coleman, the 12-hour shift is the most acceptable in the Queensland situation, given the industry requirements, physiological considerations and the isolated small town community structure. This view is based upon the goal of concentrating work and "maximizing leisure" (that is, allowing the miner the most consecutive days off so he can escape to the coast, away from the entertainment-poor mining towns). In this model of shiftwork, the individual physiology and desires of men are privileged over the physiology and routines of family and community, including school schedules and weekend social life.

run single person's quarters. Some of the more enlightened union officials are talking about "socially acceptable" roster systems, that is, ones that will accommodate the need for days off to perform family duties. They are looking at shift systems in female dominated industries for pointers. Still, an awareness of the ways in which the timing of household production, with its constant day by day demand for labor which peaks in intensified periods of activity at the beginning and end of each day, is incompatible with mining production work bunched into short bursts of extremely long days or nights, has yet to be developed. In the discussions over different roster options between companies and unions, women are still not involved.

The current conjuncture offers important political opportunities for mine workers and women alike to transform the class processes they participate in. As yet, the possibility of introducing flexible work shifts that would facilitate the emergence of more communal class processes in the home has not been put on the agenda. Yet while male miners remain content to patch up and wrestle with their marriage in the workplace to capitalist class processes, women are less content in the home. Their struggles in households might well have the potential to overdetermine struggles in capitalist worksites in ways that we have rarely imagined. In rethinking class struggles (or the seeming lack of them) in mining towns and the relationship of women to them, we have begun the task of conceptualizing new relationships between industry and household change, and perhaps suggested new ways of understanding and enacting the politics of class.

One of the goals of this chapter is to occupy a discourse of economic (and class) difference – to "naturalize" the possibility of noncapitalist economic activity by representing it as already taking place, in this case, in households. Not only is the economy portrayed as not fully capitalist, but noncapitalist activity is seen as more socially inclusive – in the sense that it involves women, men and children over a greater portion of their lives. In this discursive setting, to talk of the economy in these towns as "capitalist," or of the mining industry as "the local economy," is to ignore and do violence to the differentiations of economic space.

Yet at the same time that we have constituted a space of economic difference, we have obscured and negated economic and other sorts of differences in a number of ways. Households are represented as predominantly "feudal," in the sense that most households in these communities are seen to fall into this category, and most productive activities in each household do so as well. This homogeneous characterization of both individual households and households as a group is very different, for example, from Cameron's (1995) representation of male-female households as themselves sites of difference, characterized by multiple class processes – so that the ironing might be done by the man in an independent class process, whereas the cooking might be done by both partners in a communal class process, and the cleaning by the woman under a feudal domestic exploitative regime. We have glossed over – or, more accurately, not developed – the types of complex stories that Cameron tells, in order to develop a single story about the interaction of industrial and household class processes in coal-mining towns. This reflects our specific discursive context, political project, and rhetorical strategy: it is not the homogeneity of the household that we are trying to disrupt, but the homogeneity of economic representation. Moreover, we are attempting to demonstrate the interaction between capitalist and noncapitalist economic practices, and to emphasize the effectivity of the nonindustrial economy in the industrial domain. In our interventions in union debates over changes in work practices, we had a very specific point to make – that household and community conditions are important constituents of industrial development, which should be taken into consideration when industrial decisions are made. This theoretical argument is potentially an important ingredient in legitimating women's and community voices in the arena of union politics and industrial arbitration (an arena in which both sorts of voices have long been silent). In making this unfamiliar and perhaps unpalatable argument and attempting to liberate and legitimate unfamiliar industrial subjects, we have drawn upon representations of households and industry that are compatible with local understandings.

As with many theoretical decisions, however, ours has some unfortunate consequences. For example, at the same time that we have portrayed these single industry towns as sites of economic difference, we have represented difference as structured by two samenesses counterposed. The capitalist industry/feudal household duo is itself a familiar and traditional partnership, one that resonates with socialist feminist notions of separate capitalist spheres of production and reproduction and with the "dual systems" project of certain Marxist feminists (see chapters 2 and 3) who attempt to theorize a domestic mode of production. The first of these constitutes the economy as a (capitalist) unity, while the second restricts economic diversity to two productive systems stabilized in their respective domains.

Our dualized representation also resonates with binary and hierarchical formulations of public/private and male/female, to name just a few. But perhaps most problematically we have created a social representation in which gender, exploitation, and power are related to each other in noncontradictory ways. To the extent that the mining-town women of Central Queensland are represented as exploited and less powerful, and the men as the relatively powerful exploiters, we have produced an image that is consonant with a vision of centered subjects locked in quite obdurate social structures of "oppression" and "domination." While we might be adamant that the household situation in Central Queensland is unusual within contemporary Australia, and is therefore not to be seen as representative of male–female household relations, we have nevertheless produced yet another representation of "female exploitation and male domination."

With respect to the potential "performativity" of the chapter, it seems that in some ways we have subverted the intentions of the book – one of which is to produce discursive figurings of economic and social relations that, in Sedgwick's terms, do not "line up" (1993: 6). In pursuing this "politics of performative representation" we are attempting to speak to political subjects who both inhabit and imagine a social world that is unstructured by simple correspondences and invariances, yet complexly resistant to the will. Often we see this project as antagonistic to the political project of delineating a hegemonic formation where things line up and fall together. Portrayals of hegemony and domination have historically been instrumental in developing political subjects with a will to set history on a different course, or render society a different sort of place. But we feel slightly fearful of the form of politics that is motivated through recruitment to hegemonic stories or social representations. Such narratives and images, when it comes to class politics, are now almost invariably linked with intimations of defeat. (When it comes to the politics of gender, they are more often associated with a failure to acknowledge the history of success.) Perhaps one might say that we are interested in promoting an anti-pessimism of the intellect, as a condition of the reinvigoration of the will.

10

Haunting Capitalism: Ghosts on a Blackboard

The specter in Derrida's *Specters of Marx* is a figure of mixture and contamination, of undecidability, a reminder of the impossibility of pure and definitive being. Neither living nor dead (or if dead not absent), neither fully embodied nor entirely bodiless, the specter figures in a space "between presence and non-presence" (1994: 12). It inhabits a realm of complication and difference beyond simple antagonisms and oppositions.

In the "virtual space of spectrality" (p. 11), every presence is shaped by absences, every present moment is haunted and contaminated by the past and the future. Signaling the effectivity of what is excluded and denied, the specter refuses to let us forget the reality of what is called the unreal. It calls upon us to acknowledge the contemporaneity of the past and the future, and to countenance "the non-contemporaneity with itself of the living present" (p. xix).

Represented by Derrida in these familiar yet challenging terms, the specter is a figure for deconstruction. "Thinking the possibility of the specter" is thinking difference without opposition, beyond the oppositions between "real and unreal, . . . the living and the non-living, being and non-being." It is thinking about "possibility" itself (p. 12), especially about the actual power and powerful actuality of possibility. Derrida identifies this sort of thinking with the "messianic spirit" that he admires in *The Communist Manifesto*, where Marx and Engels invoke the specter of the future as a powerful constituent of the present moment: "a spectre is haunting Europe – the spectre of Communism."

But if Marx's messianic spirit speaks to and through Derrida, as a revenant that reminds us of future possibility (and of the future as possibility), Derrida also talks back to Marx – in the spirit of deconstruction. (He speaks respectfully of Marx as "critical but predeconstructive" [p. 170].)[1] In Derrida's reading, Marx often tries to banish the specter, to distinguish definitively between what is real and unreal.[2] His discussion of commodity fetishism, for example, in Volume 1, chapter 1 of *Capital*, attempts to assert that the material reality of a "definite social relation between men" lies behind the "fantastic form of a relation between things." This move of critical exorcism – Derrida calls it "ontologizing" but it might also be called essentialism – imbues one aspect of the commodity (such as use value, or social labor) with authentic Being and sets it in opposition to another aspect (such as exchange value, or the interaction with other commodities in commodity markets) that is less authentic or less real.[3] Yet, Derrida asks, how is it possible to isolate a pure and determinate use value entirely beyond or outside the possibility of exchange, of substitution? Marx himself acknowledges that the use value of a thing can be seen as allowing it to become an exchange value (its concrete specificities render it desirable, its very singularity is what enables its equivalence or substitutability) and that the moment of exchange is what permits its use, and therefore its coming into existence as a use value (p. 161). What does it mean to locate one aspect of the commodity's existence in the despised conceptual space of fetishism and

1 In the wake of *Specters of Marx* a number of prominent left theorists have engaged with the book (see, for example, Jameson 1995; Laclau 1995; Spivak 1995). While each of these writers has important and different points to make, they evince a shared sense of relief that Derrida has finally "come out" and acknowledged his respect for and "debt" to Marx.

2 Laclau discerns this motive in Marx's millennial political project: "Marx attempted the critique of the hauntological from the perspective of an ontology . . . the arrival at a time that is no longer "out of joint," the realization of a society fully reconciled with itself will open the way to the "end of ideology" – that is, to a purely "ontological" society which, after the consummation of the proletarian millennium, will look to hauntology as its past" (1995: 88).

3 While such rhetorical moves are often condemned out of hand as essentialist, or interpreted in such a way as to square them with an anti-essentialist reading, Garnett (1995) echoing Parker (1985) argues that the coexistence of both "modernist" and "postmodernist" elements in Marx's texts can also be read positively, as productive of certain political effects and appropriate to certain discursive contexts and projects (rather than subjected to negative readings as a sign of lapses, contradictions, and lamentable inconsistencies in Marx's thinking). "What's the problem with ontologizing or essentializing?" is, in this view, a question to which there is no generic answer but rather one that deserves a specific answer in particular discursive and political settings.

religion, and to place the other in the solid material realm of social and laboring relations? What can this sort of ontologizing achieve, except to construct a realm of authenticity in which we must ground our thinking and political projects, while devaluing other aspects of the social and cultural world? For Derrida, Marx's attempts to delimit the real are haunted by their similarity to all other such attempts. They open and leave open the unanswerable question of "how to distinguish between the analysis that denounces magic and the counter-magic that it still risks being" (p. 47).

Like his adversaries, Derrida's Marx does not want to believe in ghosts "but he thinks of nothing else . . . He believes he can oppose them, like life to death, like vain appearances of the simulacrum to real presence" (p. 47). When Marx attempts to banish the specter, in that same moment he sets himself up for a haunting – by all that must be erased, denied, cast out, mocked as chimerical or belittled as inconsequential, in order to delimit a certain objectivity.[4] Indeed the attempt to banish the specter creates the possibility and the likelihood of a haunting. In the very moment of exorcism, the specter is named and invoked, the ghost is called to inhabit the space of its desired absence. The more one attempts to render it invisible, the more spectacular its invisibility becomes.[5]

Derrida refers at length to contemporary attempts to banish the ghosts of Marx and Marxism and especially to *The End of History and the Last Man* by Francis Fukuyama, which celebrates the death of Marx and Marxism after 1989. The triumphalist text of Fukuyama is itself a rite of exorcism that involves "repeating in a mode of incantation that the dead man is really dead" (p. 48). But as Derrida points out it cannot definitively be said that Marxism is dead and therefore no longer present, that the messianic spirit of Marxism is banished once and for all, that the Marxism of the past is forgotten and buried while that of the future is unable to come into being. Derrida detects a great fear of the specter animating the triumphant eschatology of Fukuyama – an aura of certainty and hegemony undercut by a sense of vulnerability and trepidation, before the possibility of a new Marxism that will not be

[4] Later Marxists, for example, have been haunted by the question of ideology and its treatment as something derivative (in various senses of the word) of an economic realm that is somehow more authentic or grounding. For an insightful discussion of the "heterogeneous" treatments of ideology in the work of Althusser and Marx, see Parker (1985).

[5] See Keenan (1993) for a detailed discussion of the ghostly and monstrous qualities Marx cannot banish from his analysis of abstract labor (and of commodities) in the famous first chapter of *Capital*, Volume 1.

recognizable by any of the old signs (p. 50): "a ghost never dies, it remains always to come and to come-back" (p. 99).[6] For Derrida, that ghost is the hope of the future, for Fukuyama it is a prospect to be banished or suppressed.

The blackboard and the real

As a counter to the positive image of a new world order that Fukuyama sometimes holds up as an ideal and sometimes depicts as present and real, Derrida paints a "black picture on a blackboard," a quite provisional and erasable image of a world disordered and a time out of joint. On his blackboard he lists ten world conditions, including a new and ineradicable form of unemployment, homelessness on the local and international scales (with consequent exclusions from democratic participation), the war of economic competitiveness among nations, foreign debt and its companion, domestic privation, the international arms industry and trade, the spread of nuclear weapons, inter-ethnic wars, massive drug cartels – the list is enough to dispel any sense that the world is a "capitalist paradise" (p. 74) as Fukuyama and his cohort sometimes assert it to be. In Derrida's recitation of the ten plagues of "the world disorder" we can see that the characterization of the "new world order" as liberal democratic capitalism is itself an exorcism, one that cannot banish the dysfunction and destruction that it attempts both to hide and to wish away. If the world is perhaps now undeniably capitalist (for Derrida does not attempt to deny this), it is very clearly neither liberal nor democratic. This is part of what the blackboard sketch attempts to convey.

In the nature of an image on a blackboard is its provisionality, the fact that it probably won't be there tomorrow, that it is provided just for the moment of a lecture (*Specters of Marx* was initially written as a lecture). But despite its provisionality the sketch possesses (or cannot dispossess itself of) the rhetorical force of description – it makes apparent and thus to some extent renders as real and present what it describes. The blackboard sketch can be erased, yet what will the act of erasure accomplish? The terms and images of the sketch will become invisible but they will not be reworked or displaced. The provisionality of the blackboard image signals its textuality, drawing attention to the deconstructibility of its elements, but in the course of the lecture the elements and images are not

6 Derrida finds history not coming to an end but revived and repeating itself in the European Community that Fukuyama lauds as the political and economic model for the future: "As at the time of the *Manifesto*, a European alliance is formed which is haunted by what it excludes, combats or represses" (p. 61).

subjected to reinterpretation. Nor do we expect them to be. Instead they stand as provisionally real, ontologizing but erasable, deconstructible but not deconstructed, replacing and refuting the "good news" offered by Fukuyama.

The same could be said of other aspects and moments of Derrida's text, that it is descriptive and referential in parts, that it is ontological in the sense that its terms are still waiting to be deconstructed, and that its descriptions have as Derrida himself might say a performative effect – in other words, they construct or constitute reality to some undecidable extent, doing what language often does, making it difficult to see "real things" as constituted, at least in part, discursively. Here, though I am speaking generally, I am thinking particularly about that summary representation of the new world order as "liberal-democrat capitalism" (p. 80) that Derrida attributes to Fukuyama and his ilk. When human rights and democratic practices are everywhere abridged, the euphoric assertion of the triumph of liberal-democrat capitalism is for Derrida akin to the "blindest and most delirious of hallucinations" (p. 80). Yet if we read carefully this passage and what follows, the hallucination has a limit. The terms "liberal" and "democrat" turn the world upside down and inside out, shrouding it in a fog of delirium. "Capitalism," on the other hand, is not an hallucination.

If capitalism as a descriptor retains its positive status and performative force in *Specters of Marx*, this is not by virtue of its centrality to Derrida's arguments, where references to capitalism or to its metonym, the market, are usually part of the backdrop (albeit a backdrop that is quite stage-setting and efficacious). Like the deity in certain scientific treatises of the Renaissance or the Enlightenment, capitalism occupies the position of something that is not to be questioned, though the entire text provides the means and indeed the motive for calling its nature and even its existence into question.

Capitalism makes its first appearance in the opening pages of *Specters* as "capitalist imperialism," part of a familiar list of forces of victimization and oppression (including racist, nationalist and sexist violences and exterminations). Most of Derrida's other references to capitalism are similarly standard and unlikely to draw attention. Instead they occasionally bring into the foreground something that resides in a background zone of ontological presumption, functioning to guarantee the intelligibility of a particular worldview. In this sense Derrida's references to capitalism are not so much descriptions as invocations. Even when he speaks directly or at some length about capitalism (though never more than a few sentences), he draws entirely upon familiar images and descriptions. Referring, for example, to the "foreign Debt," he invokes

> the interest of capital in general, an interest that, in the order of the world today, namely the worldwide market, holds a mass of humanity under its yoke and in a new form of slavery. This happens and is authorized always in the statist or inter-statist forms of an organization. Now, these problems of the foreign Debt – and everything that is metonymized by this concept – will not be treated without at least the spirit of the Marxist critique, the critique of the market, of the multiple logics of capital, and of that which links the State and international law to this market. (p. 94)

Leaving aside certain specificities here, what is familiar (what is invoked or called up) is the image of a worldwide and conceptually monolithic market, the disembodied but no less monolithic concept of capital in general, the representation of the mass of humanity in thralldom to a singular economic structure or being, the presence of logics in the realm of economy, and the offsetting but simultaneously reinforcing vision of these logics as embedded and indeed constituted within the institutional framework of the national or supranational state. This almost ritual invocation of capital and capitalism is not a rite of exorcism – it calls forth the real rather than attempting to banish the unreal – but it has a similar "ontologizing" (to use Derrida's term once again) force. Like Marx, but much more warily and self-consciously, and without fear of ghosts, Derrida sets himself up for a haunting, one that may not be unwelcome, since he invites rather than abhors the specter.

Hauntings

> It is necessary to introduce haunting into the very construction of a concept. Of every concept, beginning with the concepts of being and time. (p. 161)

In this spirit, the spirit of Derrida, I want to pursue the ghosts that haunt the concept of capitalism. Where are the elements that contaminate the ostensibly pure and exclusively capitalist world economic space? How is its living presence shadowed by a noncapitalist past and future that are also inevitably contemporaneous though effectively suppressed? How might we characterize the heterogeneous economic spirits that wish to inhabit capitalism, to take on body within its body, within the space of the economy? What specters haunt capitalism? Certainly there must be many, some of them lurking in Derrida's own text.

Ghost No. 1

On the inspiration of Kevin St. Martin's (1995) work on cartography as constitutive of world orders, let me first call up the specter of economic difference that haunts the capitalist hegemon. St. Martin recalls the Cold War world maps of his childhood, where the countries that used to be called Communist were colored a uniform pink and appeared in the Mercator projection as a looming homogeneous mass in the northern portion of the map, stretching eastward from central Europe. The world order that was partially constituted by such maps is now seen to have collapsed, but the cartographic habits of homogeneity that characterized it are still in place. Only now it seems that the eastern bloc countries are homogeneously capitalist.

When we confront that homogeneity as an historical novelty (rather than as the logical outcome of a situation in which one half of a dualism is thought to have disappeared) it seems reasonable to ask what it might mean to call the countries of eastern Europe "capitalist." Does it mean that collective and communal and feudal and individual and family processes of production (some of which may be the same thing, and many of which co-existed with the presumptively hegemonic state sector) no longer exist? Does it mean that nonmarket exchange networks and barter systems that were in place before 1989 are no longer operative or are not now being created to deal with new problems of privation and scarcity, problems associated with a new economic and political and social order? Speaking more generally of the so-called advanced capitalist world in addition to the newly "capitalist" countries of eastern Europe, might we not argue that just as Fukuyama often poses his vision of perfected liberal democracy as a "regulating and transhistorical ideal" rather than as an historical actuality (in order to avoid certain obvious contradictions with what Derrida calls "actual history" and "realities that have an empirical appearance" [p. 62]), perfected images of capitalism and markets are themselves regulating ideals rather than literal achievements? Not only do noncapitalist practices fail to disappear over night but they cannot help but reappear (even as taints or contaminants are always appearing in anything ostensibly pure and perfected).[7]

Ghost No. 2

To get more specific about economic difference: haunting the commodity and the market are noncommodity production and nonmarket exchange

[7] What about the so-called "mixed economies" that existed in the conceptual third-space created by the duality of capitalism and communism? Are these mixtures now homogenized, purified, because "communism" no longer exists? Or has all difference become difference within (capitalism) since there is no longer difference between? If this is the case, all sorts of specters will have come forward to snatch capitalism's body.

(here we would wish, in Derrida's words, to transform "the distinction into a co-implication" [p. 161]). Often depicted as premodern and precapitalist – in other words, banished to the presumably contiguous but noncontemporaneous space of the past – these forms of production and exchange nevertheless cannot be entirely dispossessed of their contemporaneity. Noncommodity production and exchange haunt capitalism as some of its many conditions of existence, for example, noncapitalist production of goods and services in households and nonmarket exchanges both within households and within corporations.[8] The state sector in many countries is also a significant locus of noncommodity production and exchange. How are we to understand the lack of visibility and attention to these economic processes and sites? What might explain the ghostliness of noncommodity relations and transactions? Is it the absence of money that renders them colorless and de-eroticized? Is the market's force so centripetal that nonmarket practices lose meaning and consequence, becoming irrelevant and barely visible, like uncentered moments in a centered space? In the process of understanding the economy as a public arena (most often theorized as a relatively depersonalized space of monetary transactions) have the interpersonal or intraorganizational spaces of the household and the corporation become the non-economic? Here we might want to engage in the deconstructive practice of thinking difference without opposition.

Ghost No. 3

What haunts the capitalist commodity is not only *noncommodity* production (those home-cooked meals and made beds, those inputs produced internally within enterprises and transacted there) but *noncapitalist* commodity production – independent commodity production by the self-employed, slave commodities (what pure and strange moment is this, that slavery is only infrequently imagined to exist?),[9] family-based

[8] Indeed, an entire branch of economic theory (transactions costs economics) deals explicitly with the trade-offs between external or market transactions and internal or nonmarket transactions and the reasons why particular firms at particular moments might choose to engage in one or the other (see Coase 1988; North 1990; Williamson 1985).

[9] It could be seen as a moment when the voices of the "irreproachably well-bred" human rights and international law lobbies (Spivak 1995: 69) that are raised in protest against current practices of slavery (especially in the domestic service and sex industries of the Pacific Rim) often go unheard. Theorists who are conceptualizing slave relations (relations of exploitation not characterized by freedom of contract) in social sites that others have seen as capitalist include Weiner (1995).

relations of commodity production and exploitation, commodity production in collective and communal enterprises, to name just a few of the noncapitalist forms for which there are names and therefore histories. How is it that these are often confidently banished from the present – which is thereby rendered purely capitalist – or depicted as relatively marginal and inconsequential? (On this point the right and the left seem to be in some kind of imperfect but widespread agreement.)

Derrida reminds us of Marx's insistence that "the total circulation C-M-C is a 'series without beginning or end'"[10] (p. 162) and he gives this the particular reading that "metamorphosis is possible in all directions between the use-value, the commodity, and money." Another reading, not in contradiction, would acknowledge that the point of origin of circulation is not fixed or determined. Thus M-C-M′ (where money is laid out to buy commodities in order to get more money)[11] coexists with C-M-C, in other words, at any historical moment commodities are being produced and circulated both in order to expand the value of capital and to enable producers to procure other commodities. Each of these conditions (including capitalist and noncapitalist commodity production and circulation) can be seen as historically continuous and as co-existing with the other. It would appear that there is nothing "simply capitalist" about a commodity. "The market" for commodities is a space of difference, not only multiple and heterogeneous in its practices but lacking a dominant logic or relation of production.

Ghost No. 4

The concept of capitalism itself is haunted by heterogeneity, by the historicity and singularity of each form of economy that might be

[10] Here C and M refer to commodities and money respectively and the C-M-C circuit denotes the exchange of commodities for a sum of money equivalent to their value (C-M) and the subsequent purchase of commodities of value (M-C) equal to that of the commodities initially exchanged. Marx used this notation to refer to the transactions characteristic of "simple" commodity producers (as distinct from capitalists) who produce commodities in exchange for money which they use to buy subsistence and other goods and services. He also used this sequence to represent the sale of labor power within the capitalist class process: labor power (C) is exchanged for a wage (M) which is exchanged for wage goods (C).

[11] This transaction Marx associated with the merchant whose wealth is procured by buying cheaper than he sells. The increase in money (M′ less M) that accrues to the trader represents a transfer of value via a market transaction. In the very specific case of the capitalist class process the industrial capitalist engages in the transaction M-C-M′ and the increase in M is traced to the production of surplus value which is appropriated by the capitalist in the context of the purchase of labor power (LP) for a wage.

called capitalist. Each capitalist site is constituted within a social and political context, and that contextualization is itself contaminating of any pure or essential and invariant attribute associated with the concept ("it is necessary to introduce haunting into the very construction of a concept"). There is no capitalism but only capitalisms. Derrida himself acknowledges the inevitability of multiplicity and contamination when he argues parenthetically (in one of the few bracketed extensions and corrections that he added to the original text of the lecture) "with regard to capital"

> [but there is no longer, there never was just capital, nor capitalism in the singular but capitalisms plural – whether State or private, real or symbolic, always linked to spectral forces – or rather *capitalizations* whose antagonisms are irreducible.] (p. 59)

And what might this plurality entail? To take an admittedly extreme example, perhaps we could acknowledge that even the malign character of capitalism cannot be presumed. The malignancy that is its only appearance in *Specters of Marx* may not itself be free of contamination. If there are only capitalisms (and no essential capital or capitalism), some capitalist instances may be quite acceptable and benign. And if many others are malignant, for doubtless that will also be the case, it is important to ask about the contexts and conditions that produce the evil rather than accepting it as necessary and natural (for only in relation to such a question can political possibilities come to light). It might also be possible to see certain capitalist practices and institutions (some multinational corporations, for example) as relatively ineffectual and powerless, rather than as uniformly capable of dominance and self-realization.[12] We might cease to speak easily of capitalist imperialism as though empire were an aspect of capitalism's identity (albeit one that masquerades as its history).

The "last" specter

Ultimately, capitalism is haunted by its discursivity. The texts of capitalist triumphalism are interventions within what Derrida calls a "dominant discourse," one that attempts to install a particular world order and at the same time is a constituent of a current worldwide hegemonic formation. (Derrida acknowledges that his particular formulation of dominance and hegemony is itself part of his Marxian inheritance, and distances himself from the economism but not from the images of coalescence and

[12] And possibly as not uniformly capitalist. See chapter 6 on rethinking the multinational corporation.

stabilization associated with this type of Marxian social representation.) Though he takes pains to dispute Fukuyama's characterization of the new world order, he often uses the words "capital" or "capitalism" or "the market," exactly as Fukuyama does, to signify a unified economic world, one not only integrated but hegemonized by capitalism.

At the same time, Derrida calls for new conceptualizations (of politics, for example) and asks for "a profound and critical re-elaboration of the concepts of the State, of the nation-State, of national sovereignty, and of citizenship" (p. 94). Indeed, he asks us to relate to his blackboard picture not as (or not simply as) a representation of our current real distance from a regulating ideal,[13] but in the deconstructive spirit that is one of the spirits of Marxism:

> Beyond the "facts," beyond the supposed "empirical evidence," beyond all that is inadequate to the ideal, it would be a question of putting into question again, in certain of its essential predicates, the very concept of the said ideal. This would extend, for example, to the economic analysis of the market, the laws of capital, of types of capital (financial or symbolic, therefore spectral), liberal parliamentary democracy, modes of representation and suffrage, the determining content of human rights, women's and children's rights, the current concepts of equality, liberty, especially fraternity . . . dignity, the relations between man and citizen. It would also extend, in the quasi-totality of these concepts, to the concept of the human (therefore of the divine and the animal) and to a *determined* concept of the democratic that supposes it . . . (pp. 86–7)

In this passage, though it is focused on reexamining and reworking the ideal, it is clear that the spirit of deconstruction Derrida calls upon can operate to "deontologize" the economy as well. Derrida's characterization of ontology as "that which remains deconstructible"[14] situates the concepts of capital and capitalism (in his own work and elsewhere) as targets of deconstruction. If the means of deconstruction are not readily apparent, they nevertheless are available both within Marx's and within Derrida's text.

[13] This might be identified as being in the predeconstructive spirit of Marxist critique, which remains "within the idealist logic of Fukuyama" (p. 86). Yet even in this predeconstructive spirit, Derrida notes the possibility of "new modes of production" (p. 86) when he calls on Marxism to adapt and adjust its critiques to new conditions.

[14] Page 184, footnote 8.

The blackboard revisited

To return in conclusion to Derrida's blackboard, which functions not only as a provisional ontology to be deconstructed but also as a set of concepts that remain undeconstructed: I want to acknowledge it as the deployment of a certain kind of rhetorical force. I do this in the spirit of Eve Kosofsky Sedgwick's recent work on the "periperformative," which utilizes and proliferates the categories of Austinian speech act theory.[15] Sedgwick's category of the periperformative refers to utterances that are not performative in the strict sense – the strict sense being that they actually do, rather than describe the doing of, a thing ("I sentence you," "I dare you"). Periperformative acts of speech "are *about* performatives and, more properly, . . . they cluster *around* performatives" (p. 2). In Sedgwick's vision the concept of the periperformative allows us to envision potential sites of "disinterpellation" and therefore to recognize instances of the performative (she uses the specific utterance "I dare you" as her example) as "constituting a crisis in the ground or space of authority as much as it constitutes a discrete act" (p. 6). If you refuse to be dared (by invoking, for example, the periperformative "so what?") the consensual space of authority and witness is inevitably challenged and remapped. In this sense, the performative always opens the question and risk of authority.

In the case of Derrida's blackboard, we encounter the constative, but it is a reluctant constative that we find, a little like Hannah Arendt's "I have never denied being a woman or a Jew."[16] While this formulation affirms a particular ontology and participates in a particular description, it is nevertheless very different from saying "I *am* a woman and a Jew." In the sense that her statement both establishes and simultaneously intimates the real possibility of denying her womanhood and her Jewishness, Arendt creates what might be called a "crisis of reference."

Derrida's blackboard is similarly ontologizing yet differently productive of a referential crisis: one might say that it is diffidently, as opposed to negatively, referential. It seems that he has little desire to describe the world, but that also he does not wish to be caught in an appearance of denial or non-recognition of its state of dreadful weariness and disrepair. Why Marx and Marxism drive him to the world, and to the space of

15 Sedgwick (1995). Austin (1975) loosely divides speech acts into two: the constative, which describes or states; and the performative, which "does" a thing rather than describes its doing (for example, "I apologize.") In using Austin's speech act categories some philosophers (e.g., Butler) have viewed the constative as almost more successfully performative than the performative, since it constitutes or affects reality but doesn't call attention to itself in the process (Sedgwick 1995: 2).

16 Quoted in Zerilli (1995).

reference and ontology, and why they call forth in him a spirit so heavily burdened, these are considerable and very interesting questions. Ones not to be taken up here.[17]

Derrida calls attention to his constatives, the way the performative calls attention to itself. But he cannot help himself, he must utter them, he points to their deconstructibility but leaves them undeconstructed. Weary and diffident though his gesture may be, it is nonetheless forceful. It leaves us with a particular set of tasks and problems, whether political or deconstructive (not to say that these are necessarily two different things). I wish to take advantage of this diffident constative, this quasiconstative, and the force that it cannot not possess, to point to the creation here – in this chapter and indeed in this book – of a blackboard image of something other than capitalism existing and thriving on the contemporary economic scene. It's provisional and unassuming, it's clunky and unrefined – the image of noncapitalist forms of production and exchange, of noncapitalist modes of surplus labor appropriation and distribution, all those unfleshed out feudalisms, slaveries, household economic practices, intrafirm relations. But a ready option afforded by language (though undermined by deconstruction) is the possibility of ontologizing the specter. Here we do it not because we have to ("the metaphysics of presence cannot be fully evaded/expunged") or because we need to ("politics requires strategic essentialism") but perhaps because we have a great desire – to take particular advantage of the force of language, not to let the opportunity pass. What is provisional is nevertheless powerful, that's about as ontological as I want to get for now.

[17] In one who has untiringly taken on the task of destabilizing western metaphysics and philosophy, the tiredness evident in the blackboard chapter seems particularly uncharacteristic. It is interesting to think about why Derrida might have the energy to take on all of western thinking but might evince such world-weariness and lassitude in thinking about the world; one is tempted to adduce a "reality" effect, a sense of hefting the weight of the ontological, that is different from the massive but feather-light project of deconstructing philosophy. Marxism and world political economy have long been associated with, and cannot be divested of without deconstructing, the weight and gravity of "reality."

11

Waiting for the Revolution . . .

This chapter has a surplus of titles. The grand title is "Rethinking Capitalism," affirming a connection with contemporary projects to rethink received concepts and, indeed, to question the entire epistemic foundation that has rendered them prevalent and effective. The tantalizing title is "How to smash capitalism while working at home in your spare time" (this one was used at a conference hosted by *Rethinking Marxism*).[1] Last but not least there's the querulous title: "Why can feminists have revolution now, while Marxists have to wait?" This title has drawn the most criticism (since it tends to obscure the diversity within feminism and Marxism as well as the commonalities between them) but it has also provoked the greatest recognition and alignment. Despite its flippancy and falsifications, the question points to the proximity of social transformation for certain feminisms – the image of gender as something always under (re)construction, of social transformation taking place at the interpersonal level as well as the level of society as a whole. By contrast with these feminist visions, Marxism seems quite distant from both personal and social transformation.

As a Marxist I often feel envious of the feminists within and around me. My feminism reshapes the terrain of my social existence on a daily basis. Why can't my Marxism have as its object something that I am involved in (re)constructing every day? Where is my lived project of socialist construction? Certainly my sense of a socialist absence is not just a sign that Marxism is moribund while feminism, by contrast, is full of vitality. On the contrary, in academia where I am situated, Marxism appears to be thriving. It has to do, I believe, with some-

[1] Where the first version of the chapter was presented. This may explain why the chapter reads as a Marxist speaking to an audience of Marxists.

thing else – with the fact that what Marxism has been called upon to transform is something that cannot be transformed – something I will call Capitalism.[2]

Let me say this again slightly differently. Marxism has produced a discourse of Capitalism that ostensibly delineates an object of transformative class politics but that operates more powerfully to discourage and marginalize projects of class transformation.[3] In a sense, Marxism has contributed to the socialist absence through the very way in which it has theorized the capitalist presence.

Without defining Capitalism at this point, I wish to identify some of the special characteristics that give it the power to deflect socialist (and other progressive) transformations. Unlike many concepts associated with radical politics today, most prominently perhaps race, gender, and sexuality, the concept of capitalism (and by extension the concept of class, for which it is a sign in places where the term "class" cannot be used)[4] is not at the moment subject to general contestation and redefinition. Indeed there seems to be a silent consensus – within Marxism at least – that a single meaning can be associated with the word. Thus when we call the United States a capitalist country, we do so without fear of contradiction. This is not because we all have the same understanding of what capitalism is (for there may be as many capitalisms in the Marxist community as there are Marxists) but because the meaning of capitalism is not a focus of widespread rethinking and reformulation. Instead the word often functions as a touchstone, a discursive moment at which we invoke a common Marxist heritage, creating a sense of shared world views and signaling that at least *we* haven't forgotten the existence of class.

In the context of poststructuralist theory both the political subject and the social totality have been rent apart and retheorized

2 For those who have read the chapters of this book in sequence it will by now have become clear that I am referring not to "actually existing capitalism" but to prominent ways of representing capitalism within Marxist (and some nonMarxist) discourses of economy and society. To emphasize the discursive nature of my object I will, in this concluding chapter, give capitalism the respect it deserves and refer to it as Capitalism.

3 Certainly Marxism has produced many different representations of capitalism, some of which owe a substantial debt to nonMarxist theory. In this book I have constituted Capitalism as a distillation of these or, perhaps more accurately, as the residue of a filtration process that has captured certain salient elements of various Marxist theories and analyses. It is this specific residue, rather than a set of attributes common to all Marxist representations of capitalism, that I am concerned with here.

4 In certain realms of feminist thought, for example, "class is definitely non grata as a topic" but one "may creditably speak of 'proletarianization' in the context of global capitalism" (Barrett 1992: 216).

as open, continually under construction, decentered, constituted by antagonisms, fragmented, plural, multivocal, discursively as well as socially constructed. But Capitalism has been relatively immune to radical reconceptualization. Its recent development has been duly charted and tracked within the confines of traditional modernist conceptions (for example, regulation theory)[5] that have remained largely unchallenged by postmodern critical thought. Indeed, rather than being subjected to destabilization and deconstruction, Capitalism is more likely to be addressed with honorifics that evoke its powerful and entrenched position. It appears unnamed but nevertheless unmistakable as a "societal macrostructure" (Fraser and Nicholson 1990: 34), a "large-scale structure of domination" (Deutsche 1991: 19), "the global economy" or "flexible accumulation" (Harvey 1989), "post-Fordism" or even "consumer society." Often associated with an adjective that evokes its protean capacities, it emerges as "monopoly capitalism," "global capitalism," "postindustrial capitalism," "late capitalism." Like other terms of respect, these terms are seldom defined by their immediate users. Rather they function to express and constitute a shared state of admiration and subjection. For no matter how diverse we might be, how Marxist or postMarxist, how essentialist or antiessentialist, how modernist or postmodernist, most of us somewhere acknowledge that we live within something large that shows us to be small – a Capitalism, whether global or national, in the face of which all our transformative acts are ultimately inconsequential.[6]

In the representations of capitalism developed by economic theorists such as Michel Aglietta, David Harvey, Ernest Mandel, and Immanuel Wallerstein and drawn upon by a wide range of social and cultural analysts, we may see that Capitalism has a number of prominent discursive forms of appearance. I call these discursive features of Capitalism "unity," "singularity," and "totality." These features can be distinguished from each other (though none of them ever truly exists alone) and taken together (as they seldom are in particular textual settings) they constitute Capitalism as "an object of transformation that cannot be transformed." I want now to consider each of these dimensions of Capitalism in turn.

Unity

The birth of the concept of Capitalism as we know it coincided in

[5] See chapter 7.
[6] At least where capitalism is concerned.

time with the birth of "the economy" as an autonomous social sphere (Callari 1983; Poovey 1994). Not surprisingly, then, Capitalism shares with its more abstract sibling the qualities of an integrated system and the capability of reproducing itself (or of being reproduced). Like the economy, Capitalism is more often portrayed as a unified entity than as a set of practices scattered over a landscape. Represented as an organism or "system" through which flows of social labor circulate in various forms, it regulates itself according to logics or laws,[7] propelled by the life force of capital accumulation along a preordained (though not untroubled) trajectory of growth.[8]

In company with and sometimes as an alternative to organicist conceptions, the unity of Capitalism is often represented in architectural terms. Capitalism (or capitalist society) becomes a structure in which

[7] For theorists who do not wish to accord the economy the capacity to author its own causation, recognizing in this theoretical move one of the major buttresses of economic determinism (Amariglio and Callari 1989: 43) and of essentialist social theory in general (see chapter 2), the regulatory mechanisms allowing for the reproduction of capitalism may be transported outside the economy itself, so that social conditions and institutions external and contingent, rather than internal and necessary, to the capitalist economy are responsible for its maintenance and stability (see, for example, the work of the French regulation school, including Aglietta 1979, Lipietz 1987, or that of economists who theorize "social structures of accumulation," e.g., Gordon et al. 1982). Despite the expulsion of the regulatory mechanism from the economy itself, its function is unchanged, so that capitalism remains a society-wide system that has a propensity to be reproduced. Such reproductionism may characterize even hegemonic (in the Gramscian sense) conceptions of capitalism that attempt to theorize rather than presume capitalist dominance.

[8] See chapter 5. Within the organismic economy, a variety of processes may be seen as regulating capitalist reproduction and development and/or producing the integration that allows the economy to function as a unified system. The capitalist economy is seen as integrated and disciplined by the processes of the market, by competition, by the profit rate and its conditions, by the law of value or the laws of capital accumulation, all of which can be theorized as generating unity of form and movement in the economic totality. Donna Haraway notes that the functionalism inherent in organicist social conceptions has been a brake upon conceptions of the future. We are not only constrained in the present, by what the economy (here capitalism) permits and requires, but in the future, by the way its drive toward survival and self-maintenance crowds out alternative possibilities. Even when regulatory functions are externalized to dispel functionalism and attenuate economic determinism in Marxist economic discourse, the totality is still capable of being regulated (see note 7) and of being integrated and bounded in the process of regulation. Its telos is reproduction whether the mechanism guaranteeing reproduction is internal or external. Many Marxists have sidestepped the charge of functionalism by focusing on the contradictions of capitalism, but often their theories of capitalist crisis and breakdown have been imbued with an organicist conception of capitalism as a unified body/subject to life-threatening illnesses or even to death (the ultimate confirmation of organic wholeness as a form of existence).

parts are related to one another, linked to functions, and arranged "in accordance with an architecture that is internal as well as external, and no less invisible than visible" (Foucault 1973: 231).[9] The architectural/structural metaphor confers upon Capitalism qualities of durability, stability and persistence, giving it greater purchase on social reality than more ephemeral phenomena.

While Marxist conceptions usually emphasize the contradictory and crisis-ridden nature of capitalist development, capitalist crisis may itself be seen as a unifying process. Crises are often presented as originating at the organic center of a capitalist society – the relationship between capital and labor, for example, or the process of capital accumulation – and as radiating outward to destabilize the entire economic and social formation. Reconsolidation or recovery is also a process of the whole. So, for many observers of the post-World War II period, when the "long boom" ended in the crisis of Fordism, an entire Fordist "model of development" was swept aside. After a time of instability and turmoil, this society-wide structure was replaced with its post-Fordist analogue, consummating a grand economic, cultural and political realignment (see Harvey 1989; Grossberg 1992: 325–58).[10]

What is important here, for my purposes, are not the different metaphors and images of economy and society but the fact that they all confer integrity upon Capitalism. Through its architectural or organismic depiction as an edifice or body, Capitalism becomes not an uncentered aggregate of practices but a structural and systemic unity, potentially

9 Regulation theory and social structures of accumulation (SSA) theory (see note 7 above) represent two recent attempts to understand capitalism in terms of a structural model of development. Though both theoretical traditions attempt to theorize capitalist economies as the product of history and contingency rather than of logic and necessity, their analyses of particular capitalist formations conceal a structural essence of the social. This a priori and unified structure is laid bare during times of crisis and becomes particularly visible in the process of theorizing a new "model of development" or SSA, when theorists step forward to identify the new regime of accumulation, the new mode of regulation, the new labor accord, the new industrial paradigm, the new form of the state, putting flesh on society's bare bones. With the consolidation of a new model of development or SSA, the abstract and skeletal structure is once more clothed in a mantle of regulatory social practices and institutions. In this way, history is framed as a succession of analogous social structures rather than as a dynamic, contradictory and openended process that has no telos or prespecified form (see chapter 7 and Foucault 1973).

10 Interestingly, even Laclau and Mouffe (1985: 160–2) use the language of social structures of accumulation theory and regulation theory – including the term Fordism – to describe the hegemonic formation they see as structuring economic and social space in the postwar period. In doing so, they uncharacteristically fail to dissociate themselves from the a priori conceptions of social structure and totality that have accompanied these theories from their inception.

co-extensive with the national or global economy as a whole.[11] As a large, durable, and self-sustaining formation, it is relatively impervious to ordinary political and cultural interventions. It can be resisted and reformed but it cannot be replaced, except through some herculean and coordinated struggle.

Understood as a unified system or structure, Capitalism is not ultimately vulnerable to local and partial efforts at transformation. Any such efforts can always be subverted by Capitalism at another scale or in another dimension. Attempts to transform production may be seen as hopeless without control of the financial system. Socialisms in one city or in one country may be seen as undermined by Capitalism at the international scale. Capitalism cannot be chipped away at, gradually replaced or removed piecemeal. It must be transformed in its entirety or not at all.

Thus one of the effects of the unity of Capitalism is to present the left with the task of systemic transformation.

Singularity

If the unity of Capitalism confronts us with the mammoth task of systemic transformation, it is the singularity and totality of Capitalism that make the task so hopeless. Capitalism presents itself as a singularity in the sense of having no peer or equivalent, of existing in a category by itself; and also in the sense that when it appears fully realized within a particular social formation, it tends to be dominant or alone.

As a *sui generis* economic form, Capitalism has no true analogues. Slavery, independent commodity production, feudalism, socialism, primitive communism and other forms of economy all lack the systemic properties of Capitalism and the ability to reproduce and expand themselves according to internal laws.[12] Unlike socialism, for example, which is always

[11] These formulations, especially the vision of the economy as co-extensive with the nation state, attest to the overdetermination of Marxism by classical political economy and its descendants.

[12] This does not mean that these other forms have not been implicated in images of organic unity and reproducibility, for "pre-capitalist" modes of production have often been viewed as organic, stable and self-reproducing and also as revitalized by internally generated crises. But these images of organic societies have not for the most part been associated with conceptions of the economy as a special and autonomous social sphere, one that not only determines itself but by virtue of that capability tends to exert a disproportionate influence on other social locations. Moreover, when theorists of noncapitalist modes of production have attempted to conceptualize them as functioning according to laws of motion, crisis and breakdown, they have had difficulty specifying a regulatory logic with the same degree of closure as that associated with Capitalism. Theories of patriarchy as capitalism's dual have foundered on the difficulty of generating systemic laws (see chapters 2 and 3).

struggling to be born, which needs the protection and fostering of the state, which is fragile and easily deformed, Capitalism takes on its full form as a natural outcome of an internally driven growth process.

Its organic unity gives capitalism the peculiar power to regenerate itself, and even to subsume its moments of crisis as requirements of its continued growth and development. Socialism has never been endowed with that mythic capability of feeding on its own crises; its reproduction was never driven from within by a life force but always from without; it could never reproduce itself but always had to be reproduced, often an arduous if not impossible process.[13]

Other modes of production that lack the organic unity of Capitalism are more capable of being instituted or replaced incrementally and more likely to coexist with other economic forms. Capitalism, by contrast, tends to appear by itself. Thus, in the United States, if feudal or ancient classes exist, they exist as residual forms; if slavery exists, it exists as a marginal form; if socialism or communism exists, it exists as a prefigurative form. None of these forms truly and fully coexists with Capitalism. Where Capitalism does coexist with other forms, those places (the so-called Third World, for example, or backward regions in what are known as the "advanced capitalist" nations) are seen as not fully "developed." Rather than signaling the real possibility of Capitalism coexisting with noncapitalist economic forms, the coexistence of capitalism with noncapitalism marks the Third World as insufficient and incomplete. Subsumed to the hegemonic discourse of Development, it identifies a diverse array of countries as the shadowy Other of the advanced capitalist nations.

One effect of the notion of capitalist exclusivity is a monolithic conception of class, at least in the context of "advanced capitalist" countries. The term "class" usually refers to a social cleavage along the axis of capital and labor since capitalism cannot coexist with any but residual or prefigurative noncapitalist relations. The presence and fullness of the capitalist monolith not only denies the possibility of economic or class diversity in the present but prefigures a monolithic and modernist socialism – one in which everyone is a comrade and class diversity does not exist.

Capitalism's singularity operates to discourage projects to create alternative economic institutions and class relations, since these will neces-

[13] Of course, as the only true successor and worthy opponent of Capitalism, socialism is often imbued with some of Capitalism's characteristics. In order to be a suitable and commensurable replacement, for example, socialism has sometimes been theorized as having laws of motion, or a disciplinary and regulatory logic analogous to those of the market, competition, profitability and accumulation that are attributed to capitalism. But these conceptions have never become part of the dominant vision of socialism.

sarily be marginal in the context of Capitalism's exclusivity. The inability of Capitalism to coexist thus produces not only the present impossibility of alternatives but their future unlikelihood – pushing socialist projects to the distant and unrealizable future.[14]

Totality

The third characteristic of Capitalism, and perhaps its best known, is its tendency to present itself as the social totality. This is most obvious in metaphors of containment and subsumption. People who are not themselves involved in capitalist exploitation nevertheless may be seen to live "in the pores" of capitalism (Spivak 1988: 135) or within capitalism (Wallerstein 1992: 8, Grossberg 1992: 337) or under capitalism. Capitalism is presented as the embrace, the container, something large and full. Noncapitalist forms of production, such as commodity production by self-employed workers or the production of household goods and services, are seen as somehow taking place *within* capitalism. Household production becomes subsumed to capitalism as capitalist "reproduction." Even oppressions experienced along entirely different lines of social antagonism are often convened within "the plenary geography of capitalism."[15]

Capitalism not only casts a wider net than other things, it also constitutes us more fully, in a process that is more like a saturation than like a process of overdetermination. Our lives are dripping with Capitalism. We cannot get outside Capitalism; it has no outside.[16] It becomes that which has no outside by swallowing up its conditions of existence. The banking system, the national state, domestic production, the built environment, nature as product, media culture – all are conditions of Capitalism's totalizing existence that seem to lose their autonomy, their contradictory capability to be read as conditions of its nonexistence. We laboriously pry each piece loose – theorizing the legal "system," for example, as a fragmented and diverse collection of practices and institutions that is constituted by a whole host of things in addition to capitalism – but Capitalism nevertheless exerts its massive gravitational pull.

Even socialism functions as the dual or placeholder of Capitalism

[14] Those who have attempted to theorize social democracy as a transitional or mixed form of economy have encountered serious resistance from a Marxism which sees the welfare state as ultimately subsumed to or necessarily hegemonized by capitalism.

[15] Derek Gregory (1990: 81–2) commenting on Soja's (1989) treatment of "new social movements."

[16] As Gregory notes, even Laclau and Mouffe (1985) "have no difficulty recognizing that 'there is practically no domain of individual or collective life which [now] escapes capitalist relations'" (1990: 82).

rather than as its active and contradictory constituent. Socialism is just Capitalism's opposite, a great emptiness on the other side of a membrane, a social space where the fullness of Capitalism is negated. When the socialist bubble in eastern Europe burst, Capitalism flooded in like a miasma. We are all capitalist now.

It seems we have banished economic determinism and the economistic conception of class as the major axis of social transformation, only to have enshrined the economy once again – this time in a vast metonymic emplacement (Laclau and Mouffe 1985). Capitalism which is a name for a form of economy is invoked in every social dimension. The wealthy industrial societies are summarily characterized as capitalist social formations. On the one hand, we have taken back social life from the economy while, on the other, we have allowed it – under the name of Capitalism – to colonize the entire social space.[17]

This means that the left is not only presented with the revolutionary task of transforming the whole economy, it must replace the entire society as well. It is not surprising that there seems to be no room for a thriving and powerful noncapitalist economy, politics and culture, though it is heartening to consider that these nevertheless may exist.

Alternatives to Capitalism

I have characterized Marxism as producing a discourse of Capitalism that represents capitalism as unified, singular and total rather than as uncentered, dispersed, plural, and partial in relation to the economy and society as a whole.[18] I do not mean to present Marxism itself as a noncontradictory tradition – clearly Marxism has produced discourses with different and, in fact, opposite characteristics. But I detect the presence and potency of the discourse I call Capitalism in what it makes unimaginable: a contemporary socialism in places like the United States. What strikes me as an inability among Marxists to view our own activities as "socialist construction" is produced in part by a Marxist discourse, one in which capitalism is constituted as necessarily hegemonic by virtue of its own characteristics (in other words, not by virtue of historical processes or contingencies).

[17] So that while things that are associated with economism like class have become distinctly underprivileged, the economy is permitted to reassert itself in a new and more virulent form.

[18] Of course, this characterization presents as a single "discourse" something that could also be seen as scattered instances, tendencies, or remnants. Certainly the features of Capitalism that I have identified are prevalent but are not universal or uncontested. They may be recognizable to us all though none of them may characterize our own conceptions.

As Marxists we often struggle to define the discursive features of Capitalism as illusions or errors. We undermine images of Capitalism's structural or systemic unity. We criticize the ways in which Capitalism is allowed to spill over into noneconomic social domains. Yet even so the hegemony of Capitalism reasserts itself. It is visible, for example, in each new analysis that presents an economy as predominantly or monolithically capitalist. We may deprive Capitalism of self-generating capacities and structural integrity; we may rob it of the power to confer a fictional and fantastic wholeness upon our societies; but Capitalism still appears essentially alone. As the ultimate container within which we live, Capitalism is unable to coexist.

For all its variety, the discourse of Capitalism is so pervasive that it leaves us "embarrassingly empty-handed when trying to come up with a different view of things."[19] Perhaps under these circumstances the way to begin to break free of Capitalism is to turn its prevalent representations on their heads. What if we theorized capitalism not as something large and embracing but as something partial, as one social constituent among many? What if we expelled those conditions of existence – for example, property law – that have become absorbed within the conception of capitalism and allowed them their contradictory autonomy, to become conditions of existence not only of capitalism but of noncapitalism, to become conditions of capitalism's nonexistence? What if capitalism were not an entire system of economy or a macrostructure or a mode of production but simply one form of exploitation among many? What if the economy were not single but plural, not homogeneous but heterogeneous, not unified but fragmented? What if capitalism were a set of different practices scattered over the landscape that are (for convenience and in violation of difference) often seen as the same? If categories like subjectivity and society can undergo a radical rethinking, producing a crisis of individual and social identity where a presumed fixity previously existed, can't we give Capitalism an identity crisis as well? If we did, how might the "socialist project" itself be transformed?

The question is, how do we begin to see this monolithic and homogeneous Capitalism not as our "reality" but as a fantasy of wholeness, one that operates to obscure diversity and disunity in the economy and society alike?[20] In order to begin to do this we may need to get closer to redefining capitalism for ourselves. Yet this is a very difficult thing to do.[21]

If we divorce Capitalism from unity, from singularity, from totality, we are left with "capitalism" – and what might that be? Let us start where

[19] Arturo Escobar (1992: 414) speaking of the attempt to generate alternatives to the dominant discourse of Development.

most people are starting today. One of the things that has produced the sense of capitalism's ubiquity is its identification with the market, a prevalent identification outside Marxism and within Marxism one that is surprisingly not uncommon. And yet of course so many economic transactions are nonmarket transactions, so many goods and services are not produced as commodities, that it is apparent once we begin to think about it that to define capitalism as coextensive with the market is to define much economic activity as noncapitalist.

In this regard, what has for me cast the greatest light upon the discourse of Capitalism (and on the ways in which I have been confined within it without seeing its confines) have been studies of the household "economy" produced by Nancy Folbre (1993), Harriet Fraad et al. (1994), and others. These theorists represent the household in so-called advanced capitalist societies as a major locus of production and make the case that, in terms of both the value of output and the numbers of people involved, the household sector can hardly be called marginal. In fact, it can arguably be seen as equivalent to or more important than the capitalist sector. (Certainly more people are involved in household production than are involved in capitalist production.) We must therefore seek to understand the discursive marginalization of the household sector as a complex effect, one that is not produced as a simple reflection of the marginal and residual status of the household economy itself.

If we can grant that nonmarket transactions (both within and outside the household) account for a substantial portion of transactions and that therefore what we have blithely called a capitalist economy in the United States is certainly not wholly or even predominantly a market economy, perhaps we can also look within and behind the market to see the differences concealed there. The market, which has existed throughout time and over vast geographies, can hardly be invoked in any but the most general economic characterization. If we pull back this blanket term, it would not be surprising to see a variety of things wriggling beneath it. The question then becomes not whether "the market" obscures differences but how we want to characterize the differences under the blanket. As Marxists we might be interested in something other than the ways in which goods and services are transacted, though there is likely to be a

[20] I do not mean to suggest that questions about the ways in which we theorize the economy and society are simply a matter of wilful preference, but rather that they are matters of consequence. And the fact that we are not bound by some "objective reality" to represent the economy in a specified way does not mean that it is a simple or trivial matter to reconceptualize it, or that the economy and its processes are not themselves constitutive of their representations.

[21] Fortunately I am not the only one trying to do it. See, for example, Resnick and Wolff (1987) and McIntyre (1996).

wide variety of those. We might instead consider Marx's delineation of economic difference in terms of forms of exploitation, in other words, the specific forms in which surplus labor is produced, appropriated, and distributed – which indeed was what Marx was concerned to know and transform.

In any particular society we may find a great variety of forms of exploitation associated with production for a market – independent forms in which a self-employed producer appropriates her own surplus labor,[22] capitalist forms in which surplus value is appropriated from wage labor, collective or communal forms in which producers jointly appropriate surplus labor, slave forms in which surplus labor is appropriated from workers who do not have freedom of contract. None of these forms of class exploitation can be presumed to be marginal before we have even looked under the blanket.

Calling the economy "capitalist" denies the existence of these diverse economic and class processes, precluding economic diversity in the present and thus making it unlikely in the proximate future. But what if we could force Capitalism to withdraw from defining the economy *as a whole*? We might then see feudalisms, primitive communisms, socialisms, independent commodity production, slaveries, and of course capitalisms, as well as hitherto unspecified forms of exploitation. Defined in terms of the ways in which surplus labor is produced and appropriated, these diverse exploitations introduce diversity in the dimension of class – and at the same time they make thinkable (that is, apparently reasonable and realistic) the possibility of socialist class transformation.

None of this is to deny the power or even the prevalence of capitalism but to question the presumption of both. It is legitimate to theorize capitalist hegemony only if such hegemony is delineated in a theoretical field that allows for the possibility of the full coexistence of noncapitalist economic forms. Otherwise capitalist hegemony is a presumption, and one that is politically quite consequential.

[22] Ric McIntyre describes in a recent paper (1993: 231–3) the private economy of the state of Rhode Island, where the median establishment size is five. It is unlikely that all of these hire wage labor and participate in capitalist class relations, and highly likely that many of them are the locus of self-employment. What purpose is served by obscuring difference and calling these establishments capitalist, other than to affirm the hegemony of capitalism and the unlikely or marginal existence of anything else?

Conclusion

One of our goals as Marxists has been to produce a knowledge of capitalism. Yet as "that which is known," Capitalism has become the intimate enemy. We have uncloaked the ideologically-clothed, obscure monster, but we have installed a naked and visible monster in its place. In return for our labors of creation, the monster has robbed us of all force. We hear – and find it easy to believe – that the left is in disarray.

Part of what produces the disarray of the left is the vision of what the left is arrayed against. When capitalism is represented as a unified system coextensive with the nation or even the world, when it is portrayed as crowding out all other economic forms, when it is allowed to define entire societies, it becomes something that can only be defeated and replaced by a mass collective movement (or by a process of systemic dissolution that such a movement might assist). The revolutionary task of replacing capitalism now seems outmoded and unrealistic, yet we do not seem to have an alternative conception of class transformation to take its place. The old political economic "systems" and "structures" that call forth a vision of revolution as systemic replacement still seem to be dominant in the Marxist political imagination.

The New World Order is often represented as political fragmentation founded upon economic unification. In this vision the economy appears as the last stronghold of unity and singularity in a world of diversity and plurality. But why can't the economy be fragmented too? If we theorized it as fragmented in the United States, we could begin to see a huge state sector (incorporating a variety of forms of appropriation of surplus labor), a very large sector of self employed and family-based producers (most noncapitalist), a huge household sector (again, quite various in terms of forms of exploitation, with some households moving towards communal or collective appropriation and others operating in a traditional mode in which one adult appropriates surplus labor from another). None of these things is easy to see or to theorize as consequential in so-called capitalist social formations.

If capitalism takes up the available social space, there's no room for anything else. If capitalism cannot coexist, there's no possibility of anything else. If capitalism is large, other things appear small and inconsequential. If capitalism functions as a unity, it cannot be partially or locally replaced. My intent is to help create the discursive conditions under which socialist or other noncapitalist construction becomes a "realistic" present activity rather than a ludicrous or utopian future goal. To achieve this I must smash Capitalism and see it in a thousand pieces. I must make its unity a fantasy, visible as a denial of diversity and change.

In the absence of Capitalism, I might suggest a different object of socialist politics. Perhaps we might be able to focus some of our transformative energies on the exploitation and surplus distribution that go on around us in so many forms and in which we participate in various ways. In the household, in the so-called workplace, in the community, surplus labor is produced, appropriated, and distributed every day by ourselves and by others. Marx made these processes visible but they have been obscured by the discourse of Capitalism, with its vision of two great classes locked in millennial struggle. Compelling and powerful though it might be, this discourse does not allow for a variety of forms of exploitation and distribution or for the diversity of class positions and consciousnesses that such processes might participate in creating.

If we can divorce our ideas of class from systemic social conceptions, and simultaneously divorce our ideas of class transformation from projects of systemic transformation, we may be able to envision local and proximate socialisms. Defining socialism as the communal production, appropriation and distribution of surplus labor, we could encounter and construct it at home, at work, at large. These "thinly defined" socialisms wouldn't remake our societies overnight in some total and millennial fashion (Cullenberg 1992) but they could participate in constituting and reconstituting them on a daily basis. They wouldn't be a panacea for all the ills that we love to heap on the doorstep of Capitalism, but they could be visible and replicable now.[23]

To step outside the discourse of Capitalism, to abjure its powers and transcend the limits it has placed on socialist activity, is not to step outside Marxism as I understand it. Rather it is to divorce Marxism from one of its many and problematic marriages – the marriage to "the economy" in its holistic and self-sustaining form. This marriage has spawned a healthy lineage within the Marxist tradition and has contributed to a wide range of political movements and successes. Now I am suggesting that the marriage is no longer fruitful or, more precisely, that its recent offspring are monstrous and frail. Without delineating the innumerable grounds for bringing the marriage to an end, I would

[23] It is interesting to think about what the conditions promoting such socialisms might be, including forms of communal and collective subjectivity. Ruccio (1992) invokes notions of "community without unity" and "a community at loose ends" as well decentered and complex ideas of collectivity emerging within various left discourses of the 1990s.

like to mark its passing,[24] and to ask myself and others not to confuse its passing with the passing of Marxism itself. For Marxism directs us to consider exploitation, and that is something that has not passed away.

[24] Many Marxists will argue, rightly, that reports of the demise of Capitalism are greatly exaggerated. Likewise, Marxists, postMarxists and nonMarxists may argue that Marxism cannot be divorced from Capitalism, so many and fruitful are the progeny of this marriage and so entrenched its position and descendants. Understanding Marxism as a complex and contradictory tradition, I would say that it has room for all these positions and indeed that it always has. But I also think that space for the vision I am articulating is growing, in part because conditions external to Marxism – including certain trends within feminist thought – have allowed the anti-essentialist strain that has always existed within Marxism to gain both credibility and adherents.

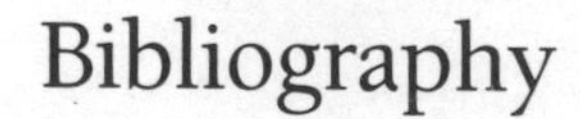

Bibliography

Aglietta, M. 1979 *A Theory of Capitalist Regulation: the U.S. experience*. Trans. D. Fernbach. London: New Left Books.

Alcorso, C. 1993 "And I'd like to thank my wife": gender dynamics and the ethnic "family business." *Australian Feminist Studies* 17: 93–108.

Althusser, L. 1969 *For Marx*. Harmondsworth: Penguin.

—— 1971 *Lenin and Philosophy and Other Essays*. London: New Left Books.

—— 1972 *Politics and History*. London: New Left Books.

—— and Balibar, E. 1970 *Reading Capital*. Trans. B. Brewster. London: Verso.

Altvater, E. 1993 *The Future of the Market: an essay on the regulation of money and nature after the collapse of "actually existing socialism."* Trans. P. Camiller. London: Verso.

Amariglio, J. 1984 "Primitive communism" and the economic development of Iroquois society. Ph.D. dissertation, University of Massachusetts–Amherst.

—— 1987 Marxism against economic science: Althusser's legacy. *Research in Political Economy* 10: 159–94.

—— 1988 The body, economic discourse and power: an economist's introduction to Foucault. *History of Political Economy* 20(4): 583–613.

—— and Callari, A. 1989 Marxian value theory and the problem of the subject: the role of commodity fetishism. *Rethinking Marxism* 2(3): 31–60.

——, Callari, A. and Cullenberg, S. 1989 Analytical Marxism: a critical overview. *Review of Social Economy* (Winter): 415–33.

——, Resnick, S. and Wolff, R. 1988 Class, power and culture. In C. Nelson and L. Grossberg (eds), *Marxism and the Interpretation of Culture*, London: Macmillan, 487–502.

—— and Ruccio, D. 1994 Postmodernism, Marxism, and the critique of modern economic thought. *Rethinking Marxism* 7(3): 7–35.

—— 1995a The (dis)orderly process of capitalist competition. In R. Bellosiore (ed.), *Marxian Economics: a centenary appraisal*, New York: Macmillan (forthcoming).

—— 1995b Modern economics and the case of the disappearing body. In M. Woodmansee and M. Osteen (eds), *The New Economic Criticism* (forthcoming).

Amin, A. (ed.) 1994 *Post-Fordism: a reader*. Oxford and Cambridge, MA: Blackwell.

Anderson, P. 1988 Modernity and revolution. In C. Nelson and L. Grossberg (eds), *Marxism and the Interpretation of Culture*, Chicago: University of Illinois Press, 317–33.

Annunziato, F. 1990 Commodity unionism. *Rethinking Marxism* 3(2): 8–33.

Armstrong, W. and McGee, T. 1985 *Theatres of Accumulation: studies in Asian and Latin American urbanisation*. London: Routledge.

Arvidson, E. 1996 An economic critique of urban planning and the "post-modern" city: Los Angeles. Ph.D. dissertation, University of Massachusetts–Amherst.

Austin, J.L. 1975 *How to Do Things with Words*. Cambridge, MA: Harvard University Press.

Australia Up-rooted. 1977 Sydney: Combined Research Center of the Amalgamated Metalworkers.

Balibar, E. 1991 *Ecrits Pour Althusser*. Paris: Edition la Decouverte.

Banuri, T. 1994 Rape as a metaphor for modernity. *Development: Journal of the Society for International Development* 1: 6–9.

Baran, P. and Sweezy, P. 1966 *Monopoly Capital*. New York: MR Press.

Barnes, T.J. 1992 Reading the texts of theoretical economic geography: the role of physical and biological metaphors. In T. J. Barnes and J. S. Duncan (eds), *Writing Worlds: discourse, text and metaphor in the representation of landscape*, London: Routledge, 118–35.

—— 1996 *Logics of Dislocation: models, methods and meanings of economic geography*. New York: Guilford Publications.

Barrett, M. 1981 *Women's Oppression Today: problems in marxist feminist analysis*. London. Verso.

—— 1988 *Women's Oppression Today: the marxist/feminist encounter*. London: Verso (second edition).

—— 1991 *The Politics of Truth: from Marx to Foucault*. Stanford, CA: Stanford University Press.

—— 1992 Words and things: materialism and method in contemporary feminist analysis. In M. Barrett and A. Phillips (eds), *Destabilizing Theory: contemporary feminist debates*, Stanford, CA: Stanford University Press, 201–19.

—— and Phillips, A. 1992 Introduction. In M. Barrett and A. Phillips (eds), *Destabilizing Theory: contemporary feminist debates*, Stanford, CA: Stanford University Press, 1–9.

Bartlett, R. 1993 *The Mabo Decision*. Sydney: Butterworth.

Battersby, C. 1993 Her body/her boundaries: gender and the metaphysics of containment. *Journal of Philosophy and the Visual Arts* 4: 30–9.

Beasley, C. 1994 *Sexual Economyths: conceiving a feminist economics*. Sydney: Allen & Unwin.

Bell, D. and Valentine, G. (eds) 1995 *Mapping Desire: geographies of sexualities*. London: Routledge.

Berger, S. 1980 Discontinuity in the politics of industrial society. In S. Berger and M. Piore (eds), *Dualism and Discontinuity in Industrial Societies*, Cambridge: Cambridge University Press, 129–41.

Best, M. 1990 *The New Competition: institutions of industrial restructuring*. Cambridge, MA: Harvard University Press.

Bhabha, H. 1990 The other question: difference, discrimination and the discourse of colonialism. In R. Ferguson, M. Gever, T. Minh-ha and C. West (eds), *Out There: marginalization and contemporary cultures*, New York: New Museum of Contemporary Art and Massachusetts Institute of Technology, 71–88.

Bhaskar, R. 1989 *Reclaiming Reality*. London: Verso.

Bina, C. and Yaghmaian, B. 1991 Post-war global accumulation and the transnationalisation of capital. *Capital and Class* 43: 107–30.

Block, F. 1990 *Postindustrial Possibilities: a critique of economic discourse*. Berkeley: University of California Press.

Bluestone, B. and Harrison, B. 1982 *The Deindustrialization of America*. New York: Basic Books.

Blunt, A. and Rose, G. (eds) 1994 *Writing Women and Space: colonial and postcolonial geographies*. New York: Guilford Press.

Bordo, S. 1989 The body and the reproduction of femininity: a feminist appropriation of Foucault. In A. Jaggar and S. Bordo (eds), *Gender Body Knowledge*, New Brunswick, NJ: Rutgers University Press, 13–34.

Boyer, R. 1990 *The Regulation School: a critical introduction*. Trans. C. Charney. New York: Columbia University.

Brown, W. 1995 *States of Injury: power and freedom in late modernity*. Princeton, New Jersey : Princeton University Press.

Bryan, D. 1992 Wages policy and the Accord: comment. *Journal of Australian Political Economy* 29: 99–110.

Buck-Morss, S. 1995 Envisioning capital: political economy on display. *Critical Inquiry* 21: 434–67.

Burden, D.S., and Googins, B.K. 1987 Boston University's balancing job and homelife study. Boston University School of Social Work, Boston, Massachusetts.

Butler, J. 1990 *Gender Trouble: feminism and the subversion of identity*. London and New York: Routledge.

—— 1995 Against proper objects. *Differences* 6 (2–3): 1–26.

Button, J. 1983 *The Steel Industry Plan News Release*. Federal Parliamentary Office of the Australian Minister for Industry and Commerce, 11 August.

Callari, A. 1983 Adam Smith, the theory of value, and the history of economic thought. Association for Economic and Social Analysis, Discussion Paper #3, University of Massachusetts–Amherst.

—— 1991 Economic subjects and the shape of politics. *Review of Radical Political Economics* 23(1–2): 201–7.

—— and Ruccio, D. (eds) 1996 *Postmodern Materialism and the Future of Marxist Theory: essays in the Althusserian tradition*. Middletown, CT: Wesleyan University Press.

Cameron, J. 1991 Women and factory employment in South East Asia: liberating the theory. Unpublished manuscript, Department of Geography, University of Sydney, NSW, Australia.

—— 1995 Ironing out the family: class, gender and power in the household. Unpublished paper, Dept. of Geography and Environmental Science, Monash University, Clayton VIC, Australia.

Campbell, B. 1986 Proletarian patriarchs and the real radicals. In V. Seddon (ed.), *The Cutting Edge: women and the pit strike*, London: Lawrence & Wishart, 249–82.

Canguilhem, G. 1988 *Ideology and Rationality in the History of the Life Sciences*. Trans. A. Goldhammer. Cambridge, MA: MIT Press.

Carter, P. 1987 *The Road to Botany Bay*. London: Faber & Faber.

Christopherson, C. 1989 Flexibility in the U.S. service economy and the emerging spatial division of labour. *Transactions of the Institute of British Geographers* N.S. 14: 131–43.

Cixous, H. 1980 The laugh of the Medusa. In E. Marks and I. de Courtivron (eds), *New French Feminisms*, Amherst: University of Massachusetts Press, 245–64.

Clark, G. 1994 Strategy and structure: corporate restructuring and the scope and characteristics of sunk costs. *Environment and Planning A* 26: 9–32.

——, Gertler, M., and Whiteman, J. 1986 *Regional Dynamics: studies in adjustment theory*. Boston: Allen & Unwin.

Coase, R. 1988 *The Firm, the Market, the Law*. Chicago: Chicago University Press.

Coffee, J.C. 1988 Shareholders versus managers: the strain in the corporate web. In J.C. Coffee, L. Lowenstein, and S. Rose-Ackerman (eds), *Knights, Raiders and Targets: the impact of the hostile takeover*, New York: Oxford University Press, 77–134.

——, Lowenstein, L. and Rose-Ackerman, S. 1988 *Knights, Raiders and Targets: the impact of the hostile takeover*. Oxford: Oxford University Press.

Cohen, S. and Zysman, J. 1987 Why manufacturing matters: the myth of the post-industrial economy. *California Management Review* 24(3): 9–26.

Coleman, R.M. 1986 *Wide Awake at 3 a.m.: by choice or by chance?* New York: W.H. Freeman and Company.

Collins, J., Alcorso, C., Castles, S., Gibson, K., and Tait. D. 1995 *A Shopful of Dreams: ethnic small business in Australia*. Sydney: Pluto.

Colomina, B. (ed.) 1992 *Sexuality and Space*. New York: Princeton Architectural Press.

Connell, R. W. 1983 *Which Way is Up?* Sydney: Allen & Unwin.

—— 1987 *Gender and Power: society, the person and sexual politics*. Cambridge: Polity Press.

—— 1995 *Masculinities*. Cambridge: Polity Press.

Cosgrove, D. 1994 Contested global visions: one-world, whole-world, and the Apollo space photographs. *Annals of the Association of American Geographers* 84(2): 270–94.

Crang, P. 1994 Displacement, consumption and identity. Paper presented at the Annual Meeting of the Association of American Geographers, San Francisco.

Cullenberg, S. 1992 Socialism's burden: toward a thin definition of socialism. *Rethinking Marxism* 5(2): 64–83.

—— 1994a Is the United States Capitalist? theoretical and empirical dimensions. Paper presented at the annual meeting of the Allied Social Sciences Association, Boston, Massachusetts, January.

—— 1994b *The Falling Rate of Profit: recasting the Marxian debate*. London: Pluto Press.

Curran, C. 1991 A change of heart: interview with Meg Smith, Peter Ewer, Chris Lloyd and John Rainford, four of the authors of *Surviving the Accord: from restraint to renewal. Australian Left Review* 134: 24–9.

Cutler, A., Hindess, B., Hirst, P., and Hussain, A. 1978 *Marx's Capital and Capitalism Today*, vol. 2. London: Routledge.

Cyert, R. M. and March, J. G. 1992 *A Behavioral Theory of the Firm*. 2nd edition. Cambridge, MA and Oxford: Blackwell.

Daley, L. 1994 Sexuate subjectivity – the infinity of two. Unpublished paper, Centre for Women's Studies, Monash University, Clayton VIC, Australia.

Daly, G. 1991 The discursive construction of economic space: logics of organization and disorganization. *Economy and Society* 20(1): 79–102.

Debord, G. 1983 *Society of the Spectacle*. Detroit: Black and Red.

Deleuze, G. and Guattari, F. 1987 *A Thousand Plateaus: capitalism and schizophrenia* Trans. B. Massumi. Minneapolis: University of Minnesota Press.

—— and Parnet, C. 1987 *Dialogues*. Trans. H. Tomlinson and B. Habberjam. New York: Columbia University Press.

Delphy, C. 1984 *Close to Home: a materialist analysis of women's oppression*. Trans. D. Leonard. London: Hutchinson.

—— and Leonard, D. 1992 *Familiar Exploitation: a new analysis of marriage in contemporary western societies*. Cambridge: Polity Press.

DeMartino, G. 1991 Trade union isolation and the catechism of the left. *Rethinking Marxism* 4 (3): 29–51.

—— 1992 Modern macroeconomic theories of cycles and crisis: a methodological critique. Ph.D. dissertation, University of Massachusetts–Amherst.

Dennis, N., Henriques, F. and Slaughter, C. 1956 *Coal is Our Life*. London: Eyre & Spottiswoode.

Derrida, J. 1978 Structure, sign and play in the discourse of the human sciences. In *Writing and Difference*. Trans. A. Bass. Chicago: University of Chicago Press, 278–93.

—— 1993 Text read at Louis Althusser's funeral. In E.A. Kaplan and M. Sprinker (eds), *The Althusserian Legacy*, London: Verso, 241–5.

—— 1994 *Specters of Marx: the state of the debt, the work of mourning, and the new international*. Trans. P. Kamuf. New York: Routledge.

Deshpande, S. and Kurtz, A. 1994 Trade tales. *Mediations* 18(1): 33–52.

Deutsche, R. 1991 Boys' Town. *Environment and Planning D: Society and Space* 9(1): 5–30.

Dicken, P. 1992 *Global Shift: the internationalization of economic activity*. New York: Guilford Press.

DiFazio, W. 1985 *Longshoremen: community and resistance on the Brooklyn waterfront*. Massachusetts: Bergin and Garvey.

Diskin, J. and Sandler, B. 1993 Essentialism and the economy in the post-Marxist imaginary: reopening the sutures. *Rethinking Marxism* 6(3): 28–48.

Dowling, R. 1993 Feminity, place and commodities: a retail case study. *Antipode* 25(4): 295–319.

Dunford, M. 1990 Theories of regulation. *Environment and Planning D: Society and Space* 8: 297–321.

Eldershaw, M. Barnard 1937 *Plaque with Laurel*. Sydney: Harrap.

Elson, D. and Pearson, R. 1981 "Nimble fingers make cheap workers": an analysis of women's employment in third world export manufacturing. *Feminist Review* 7: 87–107.

Elster, J. 1982 Marxism, functionalism and game theory. *Theory and Society* 11: 453–482.

England, K. 1991 Gender relations and the spatial structure of the city. *Geoforum* 22(2): 135–147.

Escobar, A. 1992 Reflections on "development": grassroots approaches and alternative politics. *Futures* 24 (June): 411–36.

—— 1995 *Encountering Development: the making and unmaking of the third world*. Princeton, NJ: Princeton University Press.

Eveline, J. 1993 "Bush pigs" in bulldozers: women miners and the belt shop blues. Unpublished paper, Murdoch University, Perth, WA, Australia.

Fagan, R. 1987 Australia's BHP Ltd: an emerging transnational resources corporation. *Raw Materials Report* 4: 46–55.

Fee, E. 1986 Critiques of modern science: the relationship of feminism to other radical epistemologies. In R. Bleier (ed.), *Feminist Approaches to Science*, New York: Pergamon Press, 42–56.

Ferrier, L. 1990 Mapping power: cartography and contemporary cultural theory. *Antithesis* 4(1): 35–49.

Fincher, R. 1988 Class and gender relations in the local labour market and the local state. In J. Wolch and M. Dear (eds), *The Power of Geography*, London: Unwin Hyman, 93–117.

Folbre, N. 1987 A patriarchal mode of production. In R. Albelda, C. Gunn, and W. Waller (eds), *Alternatives to Economic Orthodoxy: a reader in political economy*, Armonk, NY: M.E. Sharpe, 323–38.

—— 1993 *Who Pays for the Kids? gender and the structures of constraint*. New York and London: Routledge.

Foucault, M. 1973 *The Order of Things: an archaeology of the human sciences*. New York: Vintage Books.

—— 1980 Questions in geography. In C. Gordon (ed.), *Power/Knowledge: selected interviews and other writings 1972–1977*, New York: Random

House, 63–77.
—— 1986 Of other spaces. *Diacritics* (Spring): 22–7.
Fraad, H., Resnick, S., and Wolff, R. 1994 *Bringing It All Back Home: class, gender and power in the modern household*. London: Pluto Press.
Frank, A. G. 1969 *Capitalism and Underdevelopment in Latin America*. New York: Monthly Review Press.
Fraser, N. 1989 *Unruly Practices: power, discourse and gender in contemporary social theory*. Minneapolis, MN: University of Minnesota Press.
—— 1993 Clintonism, welfare and the antisocial wage: the emergence of a neoliberal political imaginary. *Rethinking Marxism* 6(1): 9–23.
—— and Gordon, L. 1993 A genealogy of "dependency": a keyword of the U.S. welfare state. *Signs* 19(2): 309–36.
—— and Nicholson, L. J. 1990 Social criticism without philosophy: an encounter between feminism and postmodernism. In L. J. Nicholson (ed.), *Feminism/Postmodernism*, New York: Routledge, 19–38.
Freud, S. 1930 *Civilization and its Discontents*. Trans. J. Riviere. New York: J. Cape and H. Smith.
Fröbel, F., Heinrichs, J., and Kreye, O. 1980 *The New International Division of Labor*. New York: Cambridge University Press.
Fukuyama, F. 1992 *The End of History and the Last Man*. New York: The Free Press.
Fuss, D. 1989 *Essentially Speaking: feminism, nature and difference*. New York and London: Routledge.
Gabriel, S. 1990 Ancients: a marxian theory of self-exploitation. *Rethinking Marxism* 3(1): 85–106.
Garnett, R. 1995 Marx's value theory: modern or postmodern? *Rethinking Marxism* 8,3 (forthcoming).
Gatens, M. 1991 Representations in/and the body politic. In R. Diprose and R. Ferrel (eds), *Cartographies: the mapping of bodies and spaces*. Sydney: Allen & Unwin, 79–87.
Gibson, K. 1990 Australian coal in the global context: a paradox of crisis and efficiency. *Environment and Planning A* 22: 629–46.
—— 1991a Considerations on Northern Marxist geography. *Australian Geographer* 22(1): 75–81.
—— 1991b Company towns and class processes: a study of Queensland's new coalfields. *Environment and Planning D: Society and Space* 9(3): 285-308.
—— 1993 *Different Merry-Go-Rounds: families, communities and the 7–day roster*. Brisbane: Miners Federation, Queensland Branch.
Gibson-Graham, J.K. 1994a Reflections on regions, the White Paper, and a new class politics of distribution. *Australian Geographer* 25 (2): 148–53.
—— 1994b "Stuffed if I know!": reflections on post-modern feminist social research. *Gender, Place and Culture* 1(2): 205–24.
—— 1995a Beyond patriarchy and capitalism: reflections on political subjectivity. In B. Caine and R. Pringle (eds), *Transitions: new Australian feminisms*, Sydney: Allen & Unwin, 172–83.

—— 1995b Identity and economic plurality: rethinking capitalism and "capitalist hegemony." *Environment and Planning D: Society and Space* 13: 275–82.

——, Resnick, S. and Wolff, R. (eds) 1997 *Class: The Next Postmodern Frontier*. New York: Guilford (forthcoming).

Gordon, D. 1988 The global economy: new edifice or crumbling foundations? *New Left Review* 168: 24–6.

——, Edwards, R. and Reich, M. 1982 *Segmented Work, Divided Workers*. Cambridge: Cambridge University Press.

Gossy, M. 1994 Gals and Dolls: playing with some lesbian pornography. *Art Papers* (November and December): 21–4.

Gould, S.J. 1991 *Bully for Brontosaurus: reflections in natural history*. New York: Norton.

Graham, J. 1993 Multinational corporations and the internationalization of production: an industry perspective. In G. Epstein, J. Graham and J. Nembhard (eds), *Creating a New World Economy: forces of change and plans for action*, Philadelphia: Temple University Press, 221–41.

Granovetter, M. 1985 Economic action and social structure: the problem of embeddedness. *American Journal of Sociology* 91(3): 481–510.

—— and Swedberg, R. 1992 *The Sociology of Economic Life*. Boulder, CO: Westview Press.

Greater London Council. 1985 *London Industrial Strategy*. London: GLC.

Greenfield, P. and Graham, J. 1996 Workers, communities and industrial property: an emerging language of rights. *Employee Responsibility and Rights Journal* (forthcoming).

Gregory, D. 1990 Chinatown, part three? Soja and the missing spaces of social theory. *Environment and Planning D: Society and Space* 10(4): 393–410.

Grewal, I. and Kaplan, C. (eds) 1994 *Scattered Hegemonies: postmodernity and transnqtional feminist practices*. Minneapolis: University of Minnesota Press.

Grossberg, L. 1992 *We Gotta Get Out of this Place: popular conservatism and postmodern culture*. New York and London: Routledge.

Grosz, E. 1990a Contemporary theories of power and subjectivity. In S. Gunew (ed.), *Feminist Knowledge: critique and construct*, London and New York: Routledge, 59–120.

—— 1990b Philosophy. In S. Gunew (ed.), *Feminist Knowledge: critique and construct*, London and New York: Routledge, 147–74.

—— 1992 Bodies-cities. In B. Colomina (ed.), *Sexuality and Space*, New York: Princeton Architectural Press, 241–54.

—— 1994a Architecture from the outside. Unpublished paper, Centre for Women's Studies, Monash University, Clayton VIC, Australia.

—— 1994b *Volatile Bodies: toward a corporeal feminism*. Bloomington, IN: Indiana University Press.

—— 1995 Women, *chora*, dwelling. In S. Watson and K. Gibson (eds), *Postmodern Cities and Spaces*, Oxford and Cambridge, MA: Blackwell, 47–58.

Hall, S. and Jacques, M. (eds) 1989a *New Times: the changing face of politics in the 1990s*. London: Lawrence & Wishart.

—— 1989b. Introduction. In S. Hall and M. Jacques (eds), *New Times: the changing face of politics in the 1990s*, London: Lawrence & Wishart, 11–20.

Haraway, D.J. 1991 *Simians, Cyborgs, and Women: the reinvention of nature*. New York and London: Routledge.

Harley, J. B. 1988 Maps, knowledge and power. In D. Cosgrove and S. Daniels (eds), *The Iconography of Landscape: essays on the symbolic representation, design and use of past environments*, Cambridge: Cambridge University Press.

Harrison, B. and Bluestone, B. 1988 *The Great U-Turn: corporate restructuring and the polarizing of America*. New York: Basic Books.

Hartmann, H. 1987 Changes in women's economic and family roles in post-world war II United States. In L. Benaria and C. Stimpson (eds), *Women, Households and the Economy*, New Brunswick, NJ: Rutgers University Press, 33–64.

Harvey, D. 1969 *Explanation in Geography*. London: Arnold.

—— 1973 *Social Justice and the City*. Oxford and Cambridge, MA: Blackwell.

—— 1982 *The Limits to Capital*. Oxford: Blackwell, Chicago: University of Chicago Press.

—— 1989 *The Condition of Postmodernity*. Oxford and Cambridge, MA: Blackwell.

Hazel, V. 1994 Speaking bodies: Irigaray and the politics of voice. Unpublished paper, Centre for Women's Studies, Monash University, Clayton VIC, Australia.

Heckscher, C. 1988 *The New Unionism: employee involvement in the changing corporation*. New York: Basic Books.

Hennessy, R. 1993 *Materialist Feminism and the Politics of Discourse*. New York and London: Routledge.

Herod, A. 1995 The practice of international solidarity and the geography of the global economy. *Economic Geography* 71 (4): 341–63.

Heyzer, N. 1989. The internationalization of women's work. *Southeast Asian Journal of Social Sciences* 17(2): 25–40.

Hirst, P. and Zeitlin, J. 1991 Flexible specialization versus post-Fordism: theory, evidence and policy implications. *Economy and Society* 20(1): 1–56.

Hochschild, A. 1989 *The Second Shift*. New York: Avon Books.

hooks, b. 1992 *Black Looks: race and representation*. Boston: South End Press.

Hotch, Janet. 1994. Theories and practices of self-employment: prospects for the labor movement. M.S. thesis, Labor Relations and Research Center, University of Massachusetts–Amherst.

Howitt, R. 1994a Aborigines, bauxite and gold: land, resources and identity in a rapidly changing context. Unpublished paper presented to the Mabo and Native Titles Seminar, Macquarie University Mineral and Energy Economics Centre and Australian Mining and Petroleum Law Association, Sydney.

—— 1994b SIA, sustainability and the narratives of resource regions: Aboriginal interventions in impact stories. Unpublished paper, Department of Human Geography, School of Earth Sciences, Macquarie University, NSW, Australia.

Hudson, R. and Sadler, D. 1986 Contesting works closures in Western Europe's old industrial regions: defending place or betraying class? In A. Scott and M. Storper (eds), *Production, Work, Territory*, Boston: Allen & Unwin, 172–93.

Huxley, M. 1995 Regulating the spaces of production and reproduction in the city. Australian Housing and Urban Research Institute Restructuring Difference and Social Polarisation Working Paper Series, 20 Queen Street, Melbourne VIC Australia.

Irigaray, L. 1985 *This Sex Which is Not One*. Trans. C. Porter. Ithaca, NY: Cornell University Press.

Jameson, F. 1984 Postmodernism, or the cultural logic of late capitalism. *New Left Review* 146:53–92.

—— 1991 *Postmodernism, or the Cultural Logic of Late Capitalism*. Durham, NC: Duke University Press.

—— 1995 Marx's purloined letter. *New Left Review* 209: 75–109.

Jenson, J. 1989 "Different" but not "exceptional": Canada's permeable Fordism. *Canadian Review of Sociology and Anthropology* 26(1): 69–94.

—— 1986 Gender and reproduction: or, babies and the state. *Studies in Political Economy* 20: 9–46.

Jessop, B. 1990 Regulation theories in retrospect and prospect. *Economy and Society* 19(2): 153–216.

Johnson, L. 1990 New patriarchal economies in the Australian textile industry. *Antipode* 22: 1–32.

Kaplan, E.A. and Sprinker, M. (eds) 1993 *The Althusserian Legacy*. New York: Verso.

Katz, C. and Monk, J. (eds) 1993 *Full Circles: women over the life course*. New York and London: Routledge.

Keenan, T. 1993 The point is to (ex)change it: reading Capital rhetorically. In E. Apter and W. Pietz (eds), *Fetishism*, Ithaca, New York: Cornell University Press, 152–85.

Kelly, M. and Pratt, M. J. 1992 A Foucauldian approach to accountability. Unpublished paper, University of Waikato, Hamilton, New Zealand.

Kern, S. 1983 *The Culture of Time and Space 1880–1918*. Cambridge, MA: Harvard University Press.

Kimball, G. 1983 *The 50–50 Marriage*. Boston: Beacon Press.

Kirby, V. 1992 Addressing essentialism – thoughts on the corpo-real. Women's Studies Occasional Paper Series #5, University of Waikato, Hamilton, New Zealand.

Klodawsky, F. 1995 Putting the pieces together: exploring "identity" and politics in feminist organizing against gender violence. Paper presented at the annual meeting of the Canadian Association of Geographers, Montreal.

Koechlin, T. 1989 The globalization of investment: three critical essays. Ph.D.

dissertation, University of Massachusetts, Amherst.

Kondo, D.K. 1990 *Crafting Selves: power, gender and discourses of identity in a Japanese workplace*. Chicago: University of Chicago Press.

Krieger, N. and Fee, E. 1993 What's class got to do with it? The state of health data in the United States today. *Socialist Review* 23(1): 59–82.

Laclau, E. 1977 *Politics and Ideology in Marxist Theory*. London: New Left Books.

—— 1990 *New Reflections on the Revolution of Our Time*. London: Verso.

—— 1995 "The time is out of joint." *Diacritics* 25 (2): 86–96.

—— and Mouffe, C. 1985 *Hegemony and Socialist Strategy*. London: Verso.

Larcombe, G. 1980 The political economy of Newcastle. In Stilwell, F. with Larcombe, G., *Economic Crisis, Cities and Regions*, Sydney: Pergamon Press, 146–63.

Lash, S. and Urry, J. 1987 *The End of Organized Capitalism*. Madison, WI: University of Wisconsin Press.

—— 1994 *Economies of Signs and Space*. London: Sage.

Lechte, J. 1995 (Not) belonging in postmodern space. In S. Watson and K. Gibson (eds), *Postmodern Cities and Spaces*, Oxford and Cambridge, MA: Blackwell, 99–111.

Lefebvre, H. 1991 *The Production of Space*. Trans. D. Nicholson-Smith. Oxford and Cambridge, MA: Blackwell.

Lim, L. 1990. Women's work in export factories: the politics of a cause. In I. Tinker (ed.), *Persistent Inequalities: women and world development*, New York: Oxford University Press, 101–19.

Lipietz, A. 1985 *The Enchanted World: inflation, credit and the world crisis*. London: Verso.

—— 1986 New tendencies in the international division of labor: regimes of accumulation and modes of regulation. In A. Scott and M. Storper (eds), *Production, Work, Territory*, Boston: Allen & Unwin, 16–40.

—— 1987a *Mirages and Miracles: the crises of global Fordism*. London: Verso.

—— 1987b An alternative design for the twenty-first century. No. 8738, CEPREMAP, 142, Rue de Chevaleret, 75013 Paris.

—— 1988a Building an alternative movement in France. *Rethinking Marxism* 1(3): 80–99.

—— 1988b Reflections on a Tale: the Marxist foundations of the concepts of regulation and accumulation. *Studies in Political Economy* 26 (Summer): 7–36.

—— 1992. *Towards a New Economic Order: postfordism, ecology and democracy*. New York: Oxford University Press.

—— 1993 From Althusserianism to "regulation theory." In E.A. Kaplan and M. Sprinker (eds), *The Althusserian Legacy*, New York: Verso, 99–138.

Long, P. 1985 The women of the Colorado Fuel and Iron strike. In R. Milkman (ed.), *Women, Work and Protest: a century of US women's labor history*, Boston: Routledge and Kegan Paul, 62–84.

Lynd, S. 1987 The genesis of the idea of a community right to industrial property in Youngstown and Pittsburgh, 1977–1987. *The Journal of American History* 74: 926–58.

Mackenzie, S. 1989a Women in the city. In R. Peet and N. Thrift (eds), *New Models in Geography: the political-economy perspective*, vol. 2, London: Unwin Hyman, 109–26.

—— 1989b Restructuring the relations of work and life: women as environmental actors, feminism as geographical analysis. In A. Kobayashi and S. Mackenzie (eds), *Remaking Human Geography*, London: Unwin Hyman, 40–61.

—— and Rose, D. 1983 Industrial change, the domestic economy and home life. In J. Anderson, S. Duncan, and R. Hudson (eds), *Redundant Spaces in Cities and Regions*, London: Academic Press, 155–99.

MacNeill, A. and Burczak, T. 1991 The critique of consumerism in "The Cook, the Thief, His Wife and Her Lover." *Rethinking Marxism* 4(3): 117–24.

MacWilliam, S. 1989 Manufacturing, nationalism and democracy: a review essay. *Journal of Australian Political Economy* 24: 100–20.

Magill, K. 1994 Against critical realism. *Capital and Class* 54: 113–36.

Maguire, P. 1995 Uni lecturer calls on BHP to refund the community. *The Newcastle Herald*, March 31: 9.

Mandel, E. 1975 *Late Capitalism*. London: Verso.

Marcus, S. 1992 Fighting bodies, fighting words: a theory and politics of rape prevention. In J. Butler and J. Scott (eds), *Feminists Theorize the Political*, London and New York: Routledge, 385–403.

—— 1993 Placing *Rosemary's Baby*. *Differences: A Journal of Feminist Cultural Studies* 5(3): 121–53.

Martin, L. 1995 Enterprise bargaining favours men, says study. *The Age*, Melbourne, 19 August: 4.

Marx, K. 1973 *Grundrisse: foundations of the critique of political economy*. Trans. M. Nicolaus. New York: Random House.

—— 1977 *Capital, Vol. 1*. Trans. B. Fowkes. New York: Random House.

—— 1981 *Capital, Vol. 3*. Trans. by D. Fernbach. New York: Random House.

—— and Engels, F. 1978 Manifesto of the Communist Party. In R.C. Tucker (ed.), *The Marx–Engels Reader*, New York: Norton, 469–500.

Massey, D. 1983 Industrial restructuring as class restructuring: production decentralization and local uniqueness. *Regional Studies* 17(2): 73–89.

—— 1984 *Spatial Divisions of Labor*. New York: Methuen.

—— 1988 What is an economy anyway? In J. Allen and D. Massey (eds), *The Economy in Question*, London: Sage and Open University, 229–59.

—— 1993 Politics and space/time. *New Left Review* 196: 65–84.

Massumi, B. 1987 Translator's foreword: pleasures of philosophy. In Deleuze, G. and Guattari, F., *A Thousand Plateaus: capitalism and schizophrenia*. Trans. B. Massumi. Minneapolis: University of Minnesota Press, ix–xv.

—— 1993 *A Reader's Guide to Capitalism and Schizophrenia: deviations from Deleuze and Guattari*. Cambridge, MA: MIT Press.

Mathews, J. 1989a *Tools of Change: new technology and the democratization of work*. Sydney: Pluto Press.

—— 1989b *Age of Democracy: the politics of post-Fordism*. Melbourne: Oxford University Press.

—— 1990 Towards a new model of industry development in Australia. *Industrial Relations Working Paper Series* 78: 1–21, School of Industrial Relations and Organizational Behaviour, University of New South Wales, Kensington, NSW, Australia.

McCloskey, D. 1985 *The Rhetoric of Economics*. Madison: University of Wisconsin Press.

McDowell, L. 1990 Gender divisions in a post-Fordist era – new contradictions or the same old story? Unpublished paper, Faculty of Social Sciences, The Open University, Milton Keynes, UK.

—— 1991 Life without father and Ford: the new gender order of post-Fordism. *Transactions of the Institute of British Geographers*, n.s. 16: 400–19.

—— 1994 Working in the city: spaces of power. Paper presented at the annual meeting of the Association of American Geographers, San Francisco.

McIntyre, R. 1991a Review of *Marxism in the U.S* by Paul Buhle. *Rethinking Marxism* 4(2): 149–57.

—— 1991b The political economy and class analytics of international capital flows: US industrial capital in the 1970s and 1980s. *Capital and Class* 43: 179–202.

—— 1996 Mode of production, social formation, and uneven development, or is there capitalism in America? In A. Callari and D. Ruccio (eds), *Postmodern Materialism and the Future of Marxist Theory: essays in the Althusserian Tradition*, Middletown, CT: Wesleyan University Press, 231–53.

McMichael, P. and Myhre, D. 1991 Global regulation vs. the nation-state: agro-food systems and the new politics of capital. *Capital and Class* 43: 83–106.

Metcalfe, A. 1987 Manning the mines: organising women out of class struggle. *Australian Feminist Studies* 4: 73–96.

—— 1988 *For Freedom and Dignity: historical agency and class structures in the coalfields of NSW*. Sydney: Allen & Unwin.

—— 1991 Myths of class struggle: the metaphor of war and the misunderstanding of class. *Social Analysis* 30: 77–97.

—— 1994 Crisis in Newcastle? Restructuring industry and rewriting the past. Unpublished paper, School of Sociology, University of New South Wales, Kensington, NSW, Australia.

MEWU 1992 The Australian economy and industry development: issues and challenges for Metal Workers in the 1990s. Metals and Engineering Workers' Union discussion paper, Sydney.

Miller, P. 1991 Accounting innovation beyond the enterprise: problematizing investment decisions and programming economic growth in the U.K. in the 1960s. *Accounting, Organizations and Society* 16(8): 733–62.

—— and O'Leary, T. 1987 Accounting and the construction of the governable person. *Accounting, Organizations and Society* 12(3): 235–65.

Mirowski, P. 1987 *Against Mechanism: protecting economics from science.* Totowa NJ: Rowman and Littlefield.

Mitchell, K. 1995 Flexible circulation in the Pacific Rim: capitalisms in cultural context. *Economic Geography* 71 (4): 364–82.

Mitchell, W. 1975 Wives of the radical labour movement. In A. Curthoys, S. Eade, and P. Spearitt (eds), *Women at Work*, Canberra: Australian Society for the Study of Labour History, 1–14.

Miyoshi, M. 1993 A borderless world? From colonialism to transnationalism and the decline of the nation state. *Critical Inquiry* 19 (Summer): 726–51.

Mohanty, C. T., Russo, A. and Torres, L. (eds) 1991 *Third World Women and the Politics of Feminism*. Bloomington: Indiana University Press.

Moi, T. 1985 *Sexual/Textual Politics: feminist literary theory*. London and New York: Methuen.

Montag, W. 1995 A process without a subject or goal(s): how to read Althusser's autobiography. In A. Callari., S. Cullenberg and C. Biewener (eds), *Marxism in the Postmodern Age: confronting the new world order*, New York and London: Guilford Press, 51–8.

Moodie, T.D. (with V. Ndatshe and B. Sibuyi) 1990 Migrancy and male sexuality on the South African gold mines. In M. Duberman, M. Vicinus, and G. Chauncey, Jr. (eds), *Hidden from History: reclaiming the gay and lesbian past*, New York: New American Library, 411–25.

Moon, M. 1995 Semipublics. Paper presented at the conference on Marxism and the Politics of Antiessentialism, University of Massachusetts–Amherst.

—— and Sedgwick, E. K. 1993 Divinity: a dossier, a performance piece, a little understood emotion. In E. K. Sedgwick, *Tendencies*, Durham: Duke University Press, 213–51.

——, Sedgwick, E. K.. Gianni, B. and Weir, S. 1994 Queers in (single-family) space. *Assemblage* 24 (August): 30 7.

Morris, M. 1992 *Ecstasy and Economics*. Sydney: EMPress.

Mouffe, C. 1992 Feminism, citizenship, and radical democratic politics. In J. Butler and J.W. Scott (eds), *Feminists Theorize the Political*, New York: Routledge, 369–84.

—— 1995 Post-Marxism: democracy and identity. *Environment and Planning D: Society and Space* 13 (3) 259–66.

Murray, R. 1988 Life after Henry (Ford). *Marxism Today* 6 (October): 8–13.

Negri, A. 1996 Pour Althusser: notes on the evolution of the thought of the later Althusser. In A. Callari and D. Ruccio (eds), *Postmodern Materialism and the Future of Marxist Theory: essays in the Althusserian Tradition*, Middletown, CT: Wesleyan University Press, 51–68.

Neimark, M. and Tinker, T. 1986 The social construction of management control systems. *Organizations and Society* 11(4/5): 369–95.

North, D. 1990 *Institutions, Institutional Change and Economic Performance.* Cambridge: Cambridge University Press.

Norton, B. 1986 Steindl, Levine and the inner logic of accumulation: a Marxian critique. *Social Concept* 3 (2): 43–66.

—— 1988a The power axis: Bowles, Gordon and Weisskopf's theory of postwar

U.S. accumulation. *Rethinking Marxism* 1 (3): 6–43.
—— 1988b Epochs and essences: a review of Marxist long-wave and stagnation theories. *Cambridge Journal of Economics* 12(2): 203–24.
O'Neill, P. 1994 *Capital, regulation and region: restructuring and internationalisation in the Hunter Valley NSW*. Ph.D. dissertation, Department of Human Geography, School of Earth Sciences, Macquarie University, NSW, Australia.
——, Webber, M., Weller, S., Campbell, I., Fincher, R., Matwijw, P. and Williams, C. 1995 *Labour Restructuring: a study of the Labour Adjustment Package for the textile, clothing and footwear industries, Volume 1*. Report to the Office of Labour Market Adjustment, Canberra, 1–55.
Ong, A. 1987 Disassembling gender in the electronics age. *Feminist Studies* 13: 609–26.
Parker, A. 1985 Futures for Marxism: an appreciation of Althusser. *Diacritics* 15 (4): 57–72.
—— and Sedgwick, E. K. (eds) 1995 *Performativity and Performance*. London: Routledge.
Pearce, J. 1990 UMFA to examine claims that rosters cause family break-ups. *Common Cause* 55(8): 7.
Pearson, R. 1986 Female workers in the first and third worlds: the "greening" of women's labour. In K. Purcell, S. Wood, A. Waton, S. Allen (eds), *The Changing Experience of Employment*, London: Macmillan, 75–94.
Peck, J. and Tickell, A. 1994 Searching for a new institutional fix: the after-Fordist crisis and the global-local disorder. In A. Amin (ed.), *Post-Fordism: a reader*, Oxford: Blackwell, 280–315.
Petty, B. 1978 *The Petty Age*. Sydney: Wild and Woolley.
Phizacklea, A. 1990 *Unpacking the Fashion Industry*. London: Routledge.
Phongpaichit, P. 1988 Two roads to the factory: industrialisation strategies and women's employment. In B. Agarwal (ed.), *Structures of Patriarchy: the state, the community and the household*, London: Zed, 150–63.
Piore, M. and Sabel, C. 1984 *The Second Industrial Divide: possibilities for prosperity*. New York: Basic Books.
Pollert, A. 1988 Dismantling flexibility. *Capital and Class* 34: 42–75.
Poovey, M. 1994 Making a social body: British cultural formation, 1830–1864. Paper presented to the Institute for Research on Women, Rutgers University, New Brunswick NJ.
Porpora, D., Lim, M., and Prommas, U. 1989 The role of women in the international division of labour: the case of Thailand. *Development and Change* 20: 269–94.
Pratt, G., and Hanson, S. 1991 On theoretical subtlety, gender, class, and space: a reply to Huxley and Winchester. *Environment and Planning D: Society and Space* 9: 241–6.
Pred, A. and Watts, M.J. 1992 *Reworking Modernity: capitalisms and symbolic discontent*. New Brunswick, NJ: Rutgers University Press.
Pringle, R. 1988 *Secretaries Talk: sexuality, power and work*. Sydney: Allen & Unwin.

—— 1995 Destabilising patriarchy. In B. Caine and R. Pringle (eds), *Transitions: new Australian feminisms*, Sydney: Allen & Unwin, 198–211.

Probert, B. 1995 Thinking about the White Paper: problems for a working nation. *Australian Geographer* 25(2): 103–9.

Pujol, M. A. 1992 *Feminism and Anti-Feminism in Early Economic Thought*. Aldershot, England: Edward Elgar Publishing Ltd.

Queensland Coal Board. 1990 *Annual Report*. 61 Mary Street, Brisbane, Queensland, Australia.

Rand, E. 1995 *Barbie's Queer Accessories*. Durham, NC: Duke University Press.

Resnick, S. and Wolff, R. 1987 *Knowledge and Class: a Marxian critique of political economy*. Chicago: University of Chicago Press.

—— 1989 The new Marxian economics: building on Althusser's legacy. *Economies and Societies, Serie Oeconomia – PE* 11: 185–200.

—— 1994 Between state and private capitalism: what was Soviet "socialism"? *Rethinking Marxism* 7(1): 9–30.

Roberts, S. 1994 The world is whose oyster? the geopolitics of representing globalization. Paper presented to the Annual Meeting of the Association of American Geographers, San Francisco.

Roemer, J. 1982 *A General Theory of Exploitation and Class*. Cambridge MA: Harvard University Press.

—— 1994 *A Future for Socialism*. London: Verso.

Rorty, R. 1979 *Philosophy and the Mirror of Nature*. Princeton, NJ: Princeton University Press.

Rose, D. 1989 A feminist perspective on employment and gentrification: the case of Montreal. In J. Wolch and M. Dear (eds), *The Power of Geography: how territory shapes social life*, Boston: Unwin Hyman, 118–38.

Rose, G. 1993 *Feminism and Geography: the limits of geographical knowledge*. Cambridge: Polity Press.

—— 1996 As if the mirrors had bled: masculine dwelling, masculinist theory and feminist masquerade. In N. Duncan (ed.), *(Re)placings*, London: Routledge (forthcoming).

Rouse, R. 1991 Mexican migration and the social space of postmodernism. *Diaspora* 1(1): 8–23.

Rowbotham, S. 1989 A step ahead: combining economic strategy with vision. *Interlink* (February/March): 11–14.

—— 1990 Post-Fordism. *Z Magazine* 3(9): 31–6.

Ruccio, D. 1989 Fordism on a world scale: international dimensions of regulation. *Review of Radical Political Economics* 21(4): 33–53.

—— 1991 Postmodernism and economics. *Journal of Post-Keynesian Economics* 13(4): 495–510.

—— 1992 Failure of socialism, future of socialists? *Rethinking Marxism* 5(2): 7–22.

——, Resnick, S. and Wolff, R. 1991 Class beyond the nation-state. *Capital and Class* 43: 25–42.

Saegert, S. 1980 Masculine cities and feminine suburbs. *Signs* 5 (3): 96–111.

Said, E.W. 1978 *Beginnings*. Baltimore, MD: Johns Hopkins Press.
Salais, R. and Storper, M. 1992 The four "worlds" of contemporary industry. *Cambridge Journal of Economics* 16: 169–93.
Sanyal, K. 1995 Rethinking capitalist development: toward a political economy of post-colonial capitalism. Paper presented to the Department of Economics, University of Massachusetts–Amherst.
Sassen, S. 1988 *The Mobility of Labor and Capital: a study in international investment and labor flow*. Cambridge: Cambridge University Press.
Saunders, P. and Williams, P. 1986 The new conservatism: some thoughts on recent and future developments in urban studies. *Environment and Planning D: Society and Space* 4: 393–9.
Sayer, A. and Walker, R. 1992 *The New Social Economy: reworking the division of labor*. Oxford and Cambridge, MA: Blackwell.
Scott, A. 1987 The semiconductor industry in South East Asia: organization, location and the international division of labor. *Regional Studies* 21(2): 143–59.
Scott, J.W. 1988 *Gender and the Politics of History*. New York: Columbia University Press.
Sedgwick, E.K. 1985 *Between Men: English literature and male homosocial desire*. New York: Columbia University Press.
—— 1990 *Epistemology of the Closet*. Berkeley and Los Angeles: University of California Press.
—— 1993 *Tendencies*. Durham NC: Duke University Press.
—— 1995 Around the performative: periperformative vicinities in Dickens, Eliot, and James. Unpublished manuscript, Dept. of English, Duke University, Durham, North Carolina.
Singer, J.W. 1988 The reliance interest in property. *Stanford Law Review* 40(3): 614–751.
Smith, N. 1984 *Uneven Development*. Oxford: Basil Blackwell.
—— and Katz, C. 1993 Grounding metaphor: towards a spatialized politics. In S. Pile and M. Keith (eds), *Place and the Politics of Identity*, London: Routledge, 67–83.
Smith, P. 1988 Visiting the Banana Republic. In A. Ross (ed.), *Universal Abandon: the politics of postmodernism*, Minneapolis: University of Minnesota Press, 128–48.
Soja, E. 1989 *Postmodern Geographies: the reassertion of space in critical social theory*. London: Verso.
—— and Hooper, B. 1993 The space that difference makes: some notes on the geographical margins of the new cultural politics. In S. Pile and M. Keith (eds), *Place and the Politics of Identity*, London: Routledge, 183–203.
Spivak, G.C. 1988a Can the subaltern speak? In C. Nelson and L. Grossberg (eds), *Marxism and the Interpretation of Culture*, Chicago: University of Illinois Press, 271–313.
—— 1988b *In Other Worlds: essays in cultural politics*. New York: Routledge.
—— 1995 Ghostwriting. *Diacritics* 25 (2): 65–84.

—— and Plotke, D. 1995 A dialogue on democracy. *Socialist Review* 94(3): 1–22.

Sproul, C. 1993 Mastering the Other: an ecofeminist analysis of neoclassical economics. Ph.D. dissertation, University of Massachusetts–Amherst.

St. Martin, K. 1995 Changing borders, changing cartography: possibilities for intervening in the new world order. In A. Callari, S. Cullenberg and C. Biewener (eds), *Marxism in the Postmodern Age: confronting the new world order*, New York: Guilford, 459–68.

Staeheli, L. 1994 Women and the housework of politics. Paper presented at the annual meeting of the Association of American Geographers, San Francisco.

Stead, J. 1987 *Never the Same Again: women and the miners' strike 1984–85*. London: The Women's Press.

Stearns, L. and Mizruchi, M. 1993 Corporate financing: social and economic determinants. In R. Swedberg (ed.), *Explorations in Economic Sociology*, New York: Russell Sage Foundation, 279–307.

Stilwell, F. 1991 Wages policy and the Accord. *Journal of Australian Political Economy* 28: 27–53.

Storper, M. 1990 Regional "worlds of production": conventions of learning and innovation in flexible production systems of France, Italy and the USA. Unpublished paper, Graduate School of Architecture and Urban Planning, University of California, Los Angeles.

—— and Scott, A. 1989 The geographical foundations and social regulation of flexible production complexes. In J. Wolch and M. Dear (eds), *The Power of Geography: how territory shapes social life*, Boston: Allen & Unwin, 21–40.

—— and Walker, R. 1989 *The Capitalist Imperative: territory, technology and industrial growth*. New York: Basil Blackwell.

Strauch, J. 1984 Women in rural-urban circulation networks: implications for social structural change. In J. Fawcett, S. Khoo, and P. Smith (eds), *Women in the Cities of Asia: migration and urban adaptation*, Boulder: Westview Press, 60–77.

Sturmey, R. 1989 *Women and Services in Remote Company Dominated Mining Towns*. The Rural Development Centre, University of New England, Armidale NSW 2351, Australia.

Sullivan, B. 1995 Rethinking prostitution. In B. Caine and R. Pringle (eds), *Transitions: new Australian feminisms*, Sydney: Allen & Unwin, 184–97.

Surin, K. 1994 Reinventing a physiology of collective liberation: going "beyond Marx" in the Marxism(s) of Negri, Guattari, and Deleuze. *Rethinking Marxism* 7(2): 9–27.

Swanson, G. 1995 "Drunk with glitter": consuming spaces and sexual geographies. In S. Watson and K. Gibson (eds), *Postmodern Cities and Spaces*, London: Blackwell, 80–98.

Thompson, E.P. 1963 *The Making of the English Working Class*. New York: Vintage.

Thompson, G. 1986 *Economic Calculation and Policy Formation*. London: Routledge.

Thrift, N. 1987 The geography of late twentieth-century class formation. In N. Thrift and P. Williams (eds), *Class and Space: the making of urban society*. London: Routledge, 207–53.

—— 1990 The perils of the international financial system. *Environment and Planning A* 22: 1135–40.

—— and Williams, P. (eds) 1987 *Class and Space: the making of urban society*. London: Routledge.

Timpanaro, S. 1975 *On Materialism*. Trans. L. Garner. London: New Left Books.

Tribe, K. 1981 *Genealogies of Capitalism*. London: Macmillan.

Useem, M. 1993 Shareholder power and the struggle for corporate control. In R. Swedberg (ed.), *Explorations in Economic Sociology*, New York: Russell Sage Foundation, 308–34.

Valentine, G. 1992 Images of danger: women's sources of information about the spatial distribution of male violence. *Area* 1: 22–9.

Valenze, D. 1995 *The First Industrial Women*. Oxford: Oxford University Press.

van der Veen, M. 1995 Hustling in the global economy: toward a class analysis of the "traffic in women." Paper presented at the conference on Marxism and the Politics of Antiessentialism, University of Massachusetts–Amherst.

Waldinger, R. 1986 *Through the Eye of the Needle: immigrants and enterprise in New York's garment trades*. New York: New York University Press.

Walker, A. 1945 *Coal Town: a social survey of Cessnock, NSW*. Melbourne: Melbourne University Press.

Walker, R. 1985. Class, division of labour and employment in space. In D. Gregory and J. Urry (eds), *Social Relations and Spatial Structures*, London: Macmillan, 164–89.

Wallerstein, I. 1974 *The Modern World System*. New York: Academic Press.

—— 1995 Revolution as strategy and tactics of transformation. In A. Callari, S. Cullenberg and C. Biewener (eds), *Marxism and the New World Order*, New York: Guilford Press, 225–32.

Waring, M. 1988 *Counting for Nothing: what men value and what women are worth*. Sydney: Allen & Unwin Press.

Wark, M. 1994 *Virtual Geography: living with global media events*. Bloomington, IN: Indiana University Press.

Watson, S. and Gibson, K. (eds) 1995 *Postmodern Cities and Spaces*. Oxford and Cambridge, MA: Blackwell.

Weiner, R. 1995 Marx and baseball: class analytics hits the major leagues. Paper presented to the annual summer conference of the Association for Economic and Social Analysis, Raymond, Maine.

Whitford, M. 1991 *Luce Irigaray: philosophy in the feminine*. London and New York: Routledge.

Williams, C. 1981 *Open Cut: the working class in an Australian mining town*. Sydney: Allen & Unwin.

Williamson, O. 1985 *The Economic Institutions of Capitalism*. New York: Free Press.

Wilson, E. 1991 *The Sphinx in the City: urban life, the control of disorder, and women*. London: Virago Press.
Wolf, D. 1992 *Factory Daughters: gender, household dynamics and rural industrialization in Java*. Berkeley: University of California Press.
Wolff, R. and Resnick, S. 1986 Power, property and class. *Socialist Review* 16(2):97–124.
Wolpe, H. (ed) 1980 *The Articulation of Modes of Production: essays from Economy and Society*. London: Routledge and Kegan Paul.
Worpole, K. 1992 *Towns for People: transforming urban life*. Buckingham: Open University Press.
Wright, E.O. 1978 *Class, Crisis and the State*. London: New Left Books.
—— 1985 *Classes*. London: Verso.
—— 1993 Class analysis, history and emancipation. *New Left Review* 202: 15–35.
Yeats, W.B. 1963 *The Collected Poems of W.B. Yeats*. New York: Macmillan.
Young, M. 1985 *Darwin's Metaphor: nature's place in Victorian culture*. London: Cambridge University Press.
Zerilli, L. 1995 The Arendtian body. In B. Honig (ed.), *Feminist Interpretations of Hannah Arendt*, University Park PA: Pennsylvania State University Press, 167–89.

Index